S0-DUU-568

J. A. SEIGEL

SPACE SCIENCE IN THE TWENTY-FIRST CENTURY: IMPERATIVES FOR THE DECADES 1995 TO 2015

OVERVIEW

Report of the Study Steering Group
Space Science Board
Commission on Physical Sciences, Mathematics, and Resources
National Research Council

NATIONAL ACADEMY PRESS
Washington, D.C. 1988

TL
787.5
.S63
1988
V.1

NATURAL HISTORY MUSEUM OF LOS ANGELES COUNTY
library

National Academy Press • 2101 Constitution Avenue, N.W. • Washington, D. C. 20418

NOTICE: The project that is the subject of this report was approved by the Governing Board of the National Research Council, whose members are drawn from the councils of the National Academy of Sciences, the National Academy of Engineering, and the Institute of Medicine. The members of the committee responsible for the report were chosen for their special competences and with regard for appropriate balance.

This report has been reviewed by a group other than the authors according to procedures approved by a Report Review Committee consisting of members of the National Academy of Sciences, the National Academy of Engineering, and the Institute of Medicine.

The National Academy of Sciences is a private, nonprofit, self-perpetuating society of distinguished scholars engaged in scientific and engineering research, dedicated to the furtherance of science and technology and to their use for the general welfare. Upon the authority of the charter granted to it by the Congress in 1863, the Academy has a mandate that requires it to advise the federal government on scientific and technical matters. Dr. Frank Press is president of the National Academy of Sciences.

The National Academy of Engineering was established in 1964, under the charter of the National Academy of Sciences, as a parallel organization of outstanding engineers. It is autonomous in its administration and in the selection of its members, sharing with the National Academy of Sciences the responsibility for advising the federal government. The National Academy of Engineering also sponsors engineering programs aimed at meeting national needs, encourages education and research, and recognizes the superior achievements of engineers. Dr. Robert M. White is president of the National Academy of Engineering.

The Institute of Medicine was established in 1970 by the National Academy of Sciences to secure the services of eminent members of appropriate professions in the examination of policy matters pertaining to the health of the public. The Institute acts under the responsibility given to the National Academy of Sciences by its congressional charter to be an adviser to the federal government and, upon its own initiative, to identify issues of medical care, research, and education. Dr. Samuel O. Thier is president of the Institute of Medicine.

The National Research Council was organized by the National Academy of Sciences in 1916 to associate the broad community of science and technology with the Academy's purposes of furthering knowledge and advising the federal government. Functioning in accordance with general policies determined by the Academy, the Council has become the principal operating agency of both the National Academy of Sciences and the National Academy of Engineering in providing services to the government, the public, and the scientific and engineering communities. The Council is administered jointly by both Academies and the Institute of Medicine. Dr. Frank Press and Dr. Robert M. White are chairman and vice chairman, respectively, of the National Research Council.

Support for this project was provided by Contract NASW 3482 between the National Academy of Sciences and the National Aeronautics and Space Administration.

Library of Congress Catalog Card Number 87-43329

ISBN 0-309-03838-3

Printed in the United States of America

NATIONAL RESEARCH COUNCIL

2101 CONSTITUTION AVENUE WASHINGTON, D. C. 20418

OFFICE OF THE CHAIRMAN

Dr. James Fletcher
Administrator
National Aeronautics and Space Administration
Washington, DC 20546

Dear Dr. Fletcher:

I am pleased to transmit Space Science in the Twenty-First Century: Imperatives for the Decades 1995 to 2015, a report of the Space Science Board of the National Research Council.

The report represents an impressive effort by a large number of scientists whose interests and expertise span the vast extent of space science. The Board charged the participants in the project to think broadly and creatively, and the product demonstrates clearly that they took this charge to heart. The several volumes of the report present a varied and exciting picture of opportunities in the space sciences in the future.

I want to note two aspects of the report--and of the view of space sciences presented in it--both of which are considered in the document itself but which bear repeating here. Any portrayal of the future of space science presupposes successful solutions to the severe problems that our nation's space science program faces today. The Challenger accident, coupled with our over-reliance on manned launch capabilities, has, to all intents and purposes, crippled our space science program by depriving us of access to space. The lessons of the past few years are painfully clear, and it is to be hoped that they will lead to a more balanced and resilient space program in the near and longer-term future.

Particularly in light of current uncertainties, the findings and recommendations contained in these volumes probably constitute, in aggregate, a much larger space science program than can be realistically anticipated in the period of time examined in the study. While they are aware of this, the Board and study group do believe that the recommendations should be pursued at the appropriate time. There is no attempt in the report to establish priorities among the recommendations. As noted in the preface to the Overview volume, the Board felt that setting priorities "would not be appropriate at this time when we do not have the benefit of the knowledge we expect to gain from major missions now planned but not yet begun."

THE NATIONAL RESEARCH COUNCIL IS THE PRINCIPAL OPERATING AGENCY OF THE NATIONAL ACADEMY OF SCIENCES AND THE NATIONAL ACADEMY OF ENGINEERING TO SERVE GOVERNMENT AND OTHER ORGANIZATIONS.

I commend the report to you as a stimulating and challenging description of the opportunities that lie before us in the space sciences.

Yours sincerely,

Frank Press
Chairman

STEERING GROUP

Thomas M. Donahue, University of Michigan, Chairman
Don Anderson, California Institute of Technology
D. James Baker, Joint Oceanographic Institutions, Inc.
Robert Berliner, Pew Scholars Program, Yale University
Bernard Burke, Massachusetts Institute of Technology
A. G. W. Cameron, Harvard College Observatory
George Field, Center for Astrophysics
Herbert Friedman, Naval Research Laboratory
Donald Hunten, University of Arizona
Francis Johnson, University of Texas at Dallas
Robert Kretsinger, University of Virginia
Stamatios Krimigis, Applied Physics Laboratory
Eugene Levy, University of Arizona
Frank B. McDonald, NASA Headquarters
John Naugle, Chevy Chase, Maryland
Joseph M. Reynolds, The Louisiana State University
Frederick Scarf, TRW Systems Park
Scott Swisher, Michigan State University
David Usher, Cornell University
James Van Allen, University of Iowa
Rainer Weiss, Massachusetts Institute of Technology

Dean P. Kastel, *Study Director*
Ceres M. Rangos, *Secretary*

TASK GROUP ON EARTH SCIENCES

Don L. Anderson, California Institute of Technology, Chairman
D. James Baker, Joint Oceanographic Institutions, Inc.
Moustafa T. Chahine, Jet Propulsion Laboratory
Adam M. Dziewonski, Harvard University
William B. Hanson, University of Texas at Dallas
Francis S. Johnson, University of Texas at Dallas
William M. Kaula, National Oceanic and Atmospheric
Administration
Michael McElroy, Harvard University
Berrien Moore III, University of New Hampshire
Ronald G. Prinn, Massachusetts Institute of Technology
S. Ichtiaque Rasool, Ecole Normale Superieure, Paris
Roger Revelle, University of California at San Diego
Raymond G. Roble, National Center for Atmospheric Research
Donald L. Turcotte, Cornell University

Paul F. Uhlir, *Staff Officer*
Anne L. Pond, *Secretary*

TASK GROUP ON PLANETARY AND LUNAR EXPLORATION

Donald Hunten, University of Arizona, Chairman
Arden Albee, California Institute of Technology
David C. Black, NASA Headquarters
Jacques Blamont, CNES
William Boynton, University of Arizona
Robert A. Brown, Space Telescope Science Institute
A. G. W. Cameron, Center for Astrophysics
Thomas Donahue, University of Michigan
Larry W. Esposito, University of Colorado
Ronald Greeley, Arizona State University
Eugene Levy, University of Arizona
Harold Masursky, U.S. Geological Survey
David Morrison, University of Hawaii at Manoa
George Wetherill, Carnegie Institution of Washington

Paul F. Uhlir, *Staff Officer*
Anne L. Pond, *Secretary*

TASK GROUP ON SOLAR AND SPACE PHYSICS

Frederick Scarf, TRW, Chairman
Roger M. Bonnet, Agence Spatiale Europeene
Guenter E. Brueckner, Naval Research Laboratory
Alexander Dessler, Marshall Space Flight Center
Thomas Holzer, National Center for Atmospheric Research
Stamatios Krimigis, Johns Hopkins Universtiy
Louis Lanzerotti, Bell Laboratories
John Leibacher, National Solar Observatory
Robert MacQueen, National Center for Atmospheric Research
Carl E. McIlwain, University of California, San Diego
Andrew Nagy, University of Michigan
Eugene N. Parker, University of Chicago
George Paulikas, Aerospace Corporation
Christopher Russell, University of California at Los Angeles
James Van Allen, University of Iowa

Richard C. Hart, *Staff Officer*
Carmela J. Chamberlain, *Secretary*

TASK GROUP ON ASTRONOMY AND ASTROPHYSICS

Bernard Burke, Massachusetts Institute of Technology, Chairman
James Roger Angel, University of Arizona
Jacques Beckers, NOAO Advanced Development Program
Andrea Dupree, Center for Astrophysics
Carl E. Fichtel, NASA Goddard Space Flight Center
George Field, Center for Astrophysics
Riccardo Giacconi, Space Telescope Science Institute
Jonathan Grindlay, Center for Astrophysics
Martin Harwit, Cornell University
Frank Low, University of Arizona
Frank McDonald, NASA Headquarters
Dietrich Muller, University of Chicago
Minoru Oda, ISAS
Klaus Pinkau, Max-Planck Institute for Plasma Physics
Kenneth A. Pounds, University of Leicester
Irwin Shapiro, Center for Astrophysics
Susan Wyckoff, Arizona State University

Richard C. Hart, *Staff Officer*
Carmela J. Chamberlain, *Secretary*

TASK GROUP ON FUNDAMENTAL PHYSICS AND CHEMISTRY

Rainer Weiss, Massachusetts Institute of Technology,
Co-Chairman
Joseph M. Reynolds, The Louisiana State University,
Co-Chairman
Peter Bender, University of Colorado
A. L. Berlad, University of California
Russell Donnelly, University of Oregon
Freeman Dyson, The Institute of Advanced Study
William M. Fairbank, Stanford University
Robert Hofstadter, Stanford University
George Homsy, Stanford University
James Langer, University of California
John Naugle, Fairchild Space Company
Rene Pellat, CNES
Remo Ruffini, Universita di Roma
Dudley Saville, Princeton University
John Robert Schrieffer, University of California

Dean P. Kastel, *Staff Director*
Ceres M. Rangos, *Secretary*

TASK GROUP ON LIFE SCIENCES

Scott Swisher, Michigan State University, Co-Chairman
David Usher, Cornell University, Co-Chairman
Meinrat Andreae, Florida State University
Stanley Awramik, University of California, Santa Barbara
Robert Berliner, Pew Scholars Program, Yale University
William DeCampli, Stanford Medical Center
James Ferris, Rensselaer Polytechnic Institute
Robert Fowles, University of Utah
Andrew Knoll, Harvard University
Robert Kretsinger, University of Virginia
Lynn Margulis, Boston University
Raymond Murray, Michigan State University
Quentin Myrvik, Wake Forest University
John Oro, University of Houston
Tobias Owen, SUNY at Stony Brook
Donald D. Trunkey, Oregon Health Services University
G. Donald Whedon, International Shrine Hospital
David White, Florida State University
Richard J. Wurtman, Massachusetts Institute of Technology
Richard Young, MATSCO
Jay M. Goldberg, University of Chicago
Harold Klein, The University of Santa Clara

Joyce M. Purcell, *Staff Officer*
Judith L. Estep, *Secretary*

SPACE SCIENCE BOARD

Thomas M. Donahue, University of Michigan, Chairman
Philip Abelson, American Association for the Advancement of Science
Roger D. Blandford, California Institute of Technology
Larry W. Esposito, University of Colorado
Jonathan E. Grindlay, Center for Astrophysics
Donald Hall, University of Hawaii
Andrew P. Ingersoll, California Institute of Technology
William M. Kaula, NOAA
Harold Klein, The University of Santa Clara
John W. Leibacher, National Solar Observatory
Michael Mendillo, Boston University
Robert O. Pepin, University of Minnesota
Roger J. Phillips, Southern Methodist University
David Raup, University of Chicago
Christopher T. Russell, University of California, Los Angeles
Blair D. Savage, University of Wisconsin
John A. Simpson, Enrico Fermi Institute, University of Chicago
George L. Siscoe, University of California, Los Angeles
L. Dennis Smith, Purdue University
Darrell F. Strobel, Johns Hopkins University
Byron D. Tapley, University of Texas at Austin

Dean P. Kastel, *Staff Director*
Ceres M. Rangos, *Secretary*

COMMISSION ON PHYSICAL SCIENCES, MATHEMATICS, AND RESOURCES

Norman Hackerman, Robert A. Welch Foundation, Chairman
George F. Carrier, Harvard University
Dean E. Eastman, IBM Corporation
Marye Anne Fox, University of Texas
Gerhart Friedlander, Brookhaven National Laboratory
Lawrence W. Funkhouser, Chevron Corporation (retired)
Phillip A. Griffiths, Duke University
J. Ross Macdonald, University of North Carolina, Chapel Hill
Charles J. Mankin, Oklahoma Geological Survey
Perry L. McCarty, Stanford University
Jack E. Oliver, Cornell University
Jeremiah P. Ostriker, Princeton University Observatory
William D. Phillips, Mallinckrodt, Inc.
Denis J. Prager, MacArthur Foundation
David M. Raup, University of Chicago
Richard J. Reed, University of Washington
Robert E. Sievers, University of Colorado
Larry L. Smarr, National Center for Supercomputing Applications
Edward C. Stone, Jr., California Institute of Technology
Karl K. Turekian, Yale University
George W. Wetherill, Carnegie Institution of Washington
Irving Wladawsky-Berger, IBM Corporation

Raphael G. Kasper, *Executive Director*
Lawrence E. McCray, *Associate Executive Director*

Preface

Early in 1984, NASA asked the Space Science Board to undertake a study to determine the principal scientific issues that the disciplines of space science would face during the period from about 1995 to 2015. This request was made partly because NASA expected the Space Station to become available at the beginning of this period, and partly because the missions needed to implement research strategies previously developed by the various committees of the board should have been launched or their development under way by that time. A two-year study was called for. To carry out the study the board put together task groups on earth sciences, planetary and lunar exploration, solar system space physics, astronomy and astrophysics, fundamental physics and chemistry (relativistic gravitation and microgravity sciences), and life sciences. Responsibility for the study was vested in a steering group whose members consisted of task group chairmen plus other senior representatives of the space science disciplines. To the board's good fortune, distinguished scientists from many countries other than the United States participated in this study.

The task groups and the steering group held four joint study sessions beginning in the summer of 1984 and ending in January 1986. Individual task groups also scheduled workshops at other times. The steering group met from June 16 to June 20, 1986, at the Woods Hole Study Center of the National Academy of Sciences to agree on the contents of the final overview report for the study. The findings and recommendations of the study are published in seven volumes: six task group reports and this overview report of the steering group. When the study began, the steering group encouraged the task groups to be imaginative in considering new directions for their disciplines. The intent was to challenge the participants to expand their horizons and to garner as many stimulating ideas as possible for future enterprises in space science. In providing this latitude for the task groups, the steering group felt that, since it was responsible for writing the official study report, it could not be bound initially to accept all the recommendations and findings of the task groups. The task group reports, therefore, are classified as resource documents for the steering group. Happily, at the study's conclusion, the steering group was able to accept nearly all of the task group recommendations. The steering group commends the task group reports to the reader for an understanding of the challenges that confront the space sciences and the insights they promise for the next century. We gratefully acknowledge the valuable contribution made by the task group members during this intensive study period. The official findings and recommendations of the study are those to be found in the steering group's overview.

Obviously, with the delay in the space science program caused by the *Challenger* accident, the period specified in the original request by NASA (1995 to 2015) cannot be taken literally. The steering group believes that the longer term program it recommends for each discipline should logically be undertaken when the near-term programs, currently being addressed in response to the science strategies developed by the committees of the Space Science Board, have been implemented, whenever that may be and however long it takes to complete the entire agenda of science objectives. The steering group has deliberately chosen not to undertake a prioritization of its recommendations. This would not be appropriate at this time when we do not have the benefit of the knowledge we expect to gain from major missions now planned but not yet begun. We expect the committees of the Space Science Board and internal NASA advisory groups at the appropriate time to establish the science priorities and to recommend the proper pace and sequence for new space science missions.

After the study had begun, Congress mandated the formation

of a National Commission of Space to propose goals for the nation's space program during the next 20 years. The commission published its report, entitled *Pioneering the Space Frontier*, before this study was completed.

Since the expertise of our study group and its parent board is in science, it is not in our special competence to comment on the program recommended by the commission in its entirety. We certainly endorse its first major thrust: "Advancing our understanding of our planet, our solar system, and the universe," and the additional thrust of advancing technology. That part of the commission's report entitled "Advancing Science" is altogether consonant with the recommendations of this report.

Our colleagues and partners at NASA have supported our work fully. With their help we believe that we have prepared a scientific strategy for NASA to implement in the twenty-first century that will add luster to an already bright set of accomplishments.

The hardworking staff of the Space Science Board headed by Dean Kastel, staff director and study director, deserves special recognition for their steadfast support and guidance in preparing this strategy.

Thomas M. Donahue, Chairman
Space Science Board

Contents

1 INTRODUCTION 1

2 EARTH SCIENCES: A MISSION TO PLANET EARTH 5
Background, 5
Earth as a Global System, 6
Scientific Themes, 9
Recommended Program: Post-1995, 10
The Role of NASA in Earth Sciences, 12
National Coordination, 13
Conclusions, 14

3 PLANETARY AND LUNAR EXPLORATION 15
Background, 15
Goals of Planetary Exploration, 16
Achievements of Planetary Exploration, 18
Future Planetary Exploration—A Balanced Planetary Program, 20
Conclusion, 25

4 SOLAR SYSTEM SPACE PHYSICS 27
Background, 27
The Sun, Solar Processes, and Variability, 29
The Sun-Earth System, the Magnetosphere, and the Aurora, 29

The Upper Atmosphere, 30
Magnetospheres of Other Planets and Comets, 31
Connections of Solar System Space Physics to Laboratory and Astrophysical Plasmas, 31
Nature of the Field, 32
Current Flight Projects, 32
Prospective Pre-1995 Missions, 33
Recommended Program: Post-1995, 34
Conclusions, 35

5 ASTRONOMY AND ASTROPHYSICS 37
Background, 37
Major Scientific Questions, 38
The Evolution of Space Astronomy, 41
Recommended Program: Post-1995, 45
Cross-Links with Other Disciplines, 49
Conclusions, 49

6 FUNDAMENTAL PHYSICS AND CHEMISTRY: RELATIVISTIC GRAVITATION AND MICROGRAVITY SCIENCE 51
Overview, 51

A. Relativistic Gravitation, 52
Background, 52
Tests of General Relativity Theory in Weak Fields, 52
Principle of Equivalence, 53
Secular Change in the Gravitational Constant, 54
Gravitational Waves, 54
Pre-1995 Program for Relativistic Gravitation, 55
Recommended Program for Relativistic Gravitation: Post-1995, 56

B. Microgravity Science, 57
Background, 57
Observation of States in Equilibrium, 57
Observation of States Destroyed by Gravity, 58
States Far from Equilibrium, 58
Conclusions and Recommendations for Microgravity Science, 58

7 LIFE SCIENCES 60
Background, 60

Life Science Goals and Major Questions, 63
Recommended Program: Post-1995, 69

8 INTERDISCIPLINARY STUDIES 74

9 HUMAN PRESENCE IN SPACE 76
Space Age Science, 76
The Scope of Human Presence in Space, 77

10 INTERNATIONAL COOPERATION 79

11 PRECONDITIONS AND INFRASTRUCTURE 82

1
Introduction

Where there is no vision, the people perish.
Prov. 29:18

The past quarter century of space science has been extraordinarily productive. The United States has held the lead in space science during most of these years, exploring new worlds, discovering new phenomena in space, and providing new ways to observe and predict changes in the global environment. The national space science program has amply fulfilled the objective of the National Aeronautics and Space Act of 1958 to extend "human knowledge of the Earth and of phenomena in the atmosphere and in space."

Moreover, it has contributed substantially to other objectives set forth in the act, including the development of space technology, the preservation of U.S. leadership in space, and the fostering of international cooperation. National investment in space science has produced a treasure of trained people and facilities that can continue to be productive far into the future. It is a perishable treasure, however, and it is eroding rapidly with the present lack of scientific missions and the aging of academic facilities.

For the past 30 years, scientific investigation has been neither the only objective of the space program of the United States, nor even the dominant one. The Apollo project and the development of the Space Transportation System and, more recently, of the Space Station were not primarily designed to respond to requirements set by the various disciplines of space science. Instead, establishing a human presence in space and accomplishment of large engineering projects for their own sake have driven a major part of our space program since the establishment of NASA in 1958. The steering group for this study recommends that the present ordering of priorities in the national space program be changed.

The steering group proposes that, as the nation considers its future in space, the advance of science and its applications to human welfare be adopted and implemented as an objective no less central to the space program of the United States than any other, such as the capability of expanding man's presence in space. Other related activities, such as the development of space technology, should be carried out so as to maximize scientific return. The steering group believes that attaining the objectives of science can provide any degree of challenge to the development of space technology that may be desired. This will ensure that the scientific and engineering resources available are effectively utilized in the national interest, as required by the act of 1958. This same standard—obtaining the greatest scientific advance for the available resources—should prevail when determining the balance between manned and unmanned space activities as well.

Provided the nation recovers full access to space rapidly after the *Challenger* accident, a large number of scientific missions will be in space—or well along toward launch—by the last years of this century. These missions promise a rich harvest of scientific results that will significantly improve our understanding of the universe. The challenge to the space sciences is to take a bold leap forward after 1995, carrying them closer to answers for the most fundamental questions about the nature of the universe. By responding to this challenge, space science will also contribute to our understanding of the delicate ecological balance that sustains life on Earth. Together, these advances will provide deeper insight into the world and our relation to it. Applications of this knowledge will directly enhance the quality of life for all humans. Moreover, new technology developed to meet the requirements of science will have many earth-bound applications.

Someday it may be possible to launch and maintain factories in space where we can profitably manufacture exotic materials. Today, there is no way to predict whether or not this will be feasible. Only a grasp of the underlying science, developed from performance of fundamental experiments in the low-gravity environment of space, will allow us to decide whether or not these aspirations are realistic, and, if so, to select the most promising avenues for development. The steering group believes that such beneficial applications of space technology as these have their best—and perhaps only—chance to flourish if science is made the principal objective of the civilian space program.

This study focuses on large-scale scientific undertakings. There is every reason to believe that, if they are to succeed, they must be built on a solid foundation of supporting research and technology, and on such small-scale exploratory projects as the present Explorer, Observer, Spartan, and suborbital programs. Supporting research must include stable funding for vigorous theoretical and laboratory studies. It is these that will provide the framework for understanding the data obtained from scientific missions.

The past 40 years of international activities at high altitudes and in space have generated a remarkable range of scientific achievements. The twenty-first century promises to build on these achievements at an accelerating rate, provided the nation furnishes the necessary resources and does not repeat its mistake of denying itself a sturdy, redundant system of access to space. Automated spacecraft, remote sensing instruments, and manned space stations will continue to add profound insights into the nature of the universe.

This report discusses the opportunities for space science in the period from 1995 to 2015. The study on which this report is based was devoted to six subjects: (1) the planet Earth; (2) planetary and lunar exploration; (3) solar system space physics; (4) astronomy and astrophysics; (5) fundamental physics and chemistry (relativistic gravitation and microgravity science); and (6) life sciences. Each subject was developed by a separate task group and is discussed in an individual volume. Collectively, these volumes set forth the scientific opportunity that exists in space research and its applications. Exploiting this opportunity should be the paramount consideration in the national debate on the goals of the civilian space program. This overview volume contains a perspective on progress in the six disciplines of space science. It also includes the prospects for major achievements by 1995 from missions already under way or awaiting new starts. Finally, it presents a set of long-range goals for these disciplines during the first two decades of the twenty-first century.

The following pages present broad themes for future scientific pursuits and highlight some examples of high-priority missions for the turn of the century. A few recommendations are cited for each discipline to suggest how these themes might be developed.

2
Earth Sciences: A Mission to Planet Earth

BACKGROUND

We now have the technology and the incentive to move boldly forward on a "Mission to Planet Earth." The steering group calls upon the nation to implement an integrated global program of fundamental research with space-borne and earth-based instrumentation. Such a program would probe the origin, evolution, and nature of our planet, its place in our solar system, and its interaction with living things, including mankind.

For earth sciences it is particularly appropriate to focus on planning for the period from 1995 to 2015. This is because the science base of this discipline is well developed. Various observational systems have already been established, and programs extending into the last decade of this century have already been proposed. The long lead times associated with the development of spacecraft and sensors mean that recommendations adopted now will not affect current programs until at least the mid-1990s. Thus, a planning document at this time is particularly relevant.

During the past 2 or 3 years, there has been an enormous amount of planning for a study of Earth as a global system, and for an observing system to monitor global change. It is clear that such a system must be largely space-based, yet the earth-based part of the measurements is integral as well. Several recent reports have helped to set the scientific context for such global studies. These have come from the National Research Council (Committee on Earth Sciences, Committee for the International Geosphere-Biosphere Program, Space Applications Board), from NASA (Global Habitability, Earth Systems Sciences Committee), and from the International Council of Scientific Unions (Committee on Global Change). The technological context in which these studies will be carried out will depend largely on the pace of development of global observing systems. (For a general policy statement on cooperation, see Chapter 10.)

EARTH AS A GLOBAL SYSTEM

The records of the first human attempts to understand Earth are lost in antiquity, but we know that early man made exploration voyages in the Pacific, Atlantic, and Indian oceans. As early as the third century B.C. the Greeks knew that Earth was a finite globe and were able to estimate its circumference. Thus, from ancient times to the present, we have used exploration and physical reasoning to understand earth processes and to explore the Earth's place in the solar system.

Modern techniques and new integrated programs have yielded improved information about the state of the atmosphere, the ocean, and the land surface. We have been able to directly measure continental drift, and to probe Earth's crust by drilling; seismic and acoustic techniques have let us probe even deeper. In addition, we now possess improved weather forecasts and new information about agricultural conditions. Measurements of winds and waves on the ocean's surface, of ocean currents, of primary productivity in the ocean, and of the chemical constitution of the atmosphere have all added to our understanding of global systems.

Very recently, interest has focused on problems where advances could have important societal impacts. These problems include the prediction of earthquakes, volcanoes, and climatic anomalies such as El Niño, whose economic impact is measured in billions of dollars. The increase in the atmosphere of carbon dioxide and other gases that may contribute to a "greenhouse" effect has also focused attention among scientists. New tools and ideas will allow us to address such problems.

We also have much new information about the atmosphere and surface properties of the other planets that will help us in understanding our own. As we have learned about the other planets of the solar system, it has become evident that Earth is different in several remarkable ways. The blue and white of Earth contrast sharply with the red of dusty Mars, the dazzling whiteness of Venus, and the complex swirling colors of Jupiter. Continued exploration has shown other, fundamental differences between planet Earth and all other planets of the solar system. The most striking of these is that living creatures have existed on Earth for more than 3.5 billion years, evolving continuously from the simplest one-celled organism to the present diversity of life forms. In contrast, it is probable that biological activity is not—and perhaps never was—present on any of the other planets during the lifetime of the solar system.

Because liquid water is essential for life on Earth, the survival and evolution of biological organisms provide a convincing argument that Earth has always had water on its surface at a temperature to keep it liquid. Without the oceans, Earth's atmosphere would be profoundly different. For instance, we have only modest amounts of carbon dioxide in our atmosphere, thus avoiding the greenhouse effect experienced by Venus. It is believed that nearly all the carbon dioxide that has flowed from Earth's interior has been buried in ocean sediments as limestone or organic carbon. The presence of free oxygen would be impossible without the photodissociation of water and the consequent escape of hydrogen. Without the presence of oxygen, ozone would not exist in the stratosphere to shield surface life from destructive solar radiation. Most animals could not then exist, since they depend on oxygen-based metabolism.

In turn, other processes must limit these ocean effects to keep Earth habitable. Oxygen in moderate amounts is a necessity for animal life, but in higher concentrations it is toxic. If organic nutrients continued to accumulate in sediments, all nutrients would eventually return to insoluble forms. If limestone sediments continued to accumulate without a compensating inflow of carbon dioxide, photosynthesis would taper off as the carbon dioxide concentration fell.

Such a compensating inflow of carbon dioxide does, in fact, occur as part of the remarkable phenomenon of plate tectonics. This process of continual recycling of Earth's surface materials into the interior, and their reappearance in mid-ocean ridges and volcanoes, is probably essential to preserving Earth's benign environment. Moreover, motions deep in Earth's interior drive the plates and generate the magnetic field that partially shields it from the harsh environment of space.

Thus it is Earth's own inner life, together with the interactions of its unique surface phenomena, that has determined its history and our own. A convective process deep in Earth's core—fired by radioactive decay and the primordial heat of agglomeration—has joined the complex interplay of the atmosphere and oceans with the biosphere to forge the world we know. However, major questions remain unanswered. Why does the phenomenon of plate tectonics operate on Earth but not on Mars and not, perhaps, on Venus? What are the characteristics of Earth that make plate tectonic convection possible? What are the nature and rate of convection? What are the effects of changing rates of convection on atmospheric carbon dioxide concentration, and hence on climate and on the biosphere? What insights can we gain from studies of the variable magnetic field generated by Earth's interior dynamo? How do the ocean and the atmosphere interact to produce long-term climate change? What is the role of the biosphere in climate? And, finally, how does Earth work as a system?

Even the origins of life may be related to plate tectonics. We have discovered complex ecosystems around deep-sea vents in the mid-ocean ridges. In the vents' scalding water live anaerobic sulfide-oxidizing bacteria that provide the energy and organic compounds for the local animal inhabitants. This environment may have been the cradle of life on Earth, despite its inaccessibility to photosynthesis. High temperatures would have allowed rapid chemical reactions and reduced sulfur compounds for energy. The overlying water would have shielded organisms from destructive ultraviolet radiation.

Another unanswered question is the effect on Earth of asteroid and comet collisions. What has been their effect on the evolution of life? The "great dyings" in the biological record may be due to these collisions, stimulating, in turn, the rapid evolution of new life forms. A careful search for evidence of such collisions in the geologic record could throw a new light on evolutionary processes.

In more general terms, it is clear that a comprehensive study, from Earth's outer atmosphere to its inmost core, is essential to understanding the conditions for life. Advances in our ability to observe the planet both from space and from Earth itself now make such a global study possible. For example, we will soon possess computers that can model the turbulent flows typical of the oceans, the atmosphere, and molten materials. Between now and 1995 many of these earth-monitoring systems will be tested, and a number of research missions for remote sensing will be carried out. As the steering group looks to the period 1995 to 2015, it foresees the application of these results to the development of an ongoing observational system for the Earth. Understanding Earth as a complex whole will begin from such global studies.

SCIENTIFIC THEMES

Four overarching scientific themes (also called "grand themes") will guide the study of earth processes:

1. Determining the composition, structure, and dynamics of the Earth's interior and crust, and its evolution.
2. Establishing and understanding the structure, dynamics, and chemistry of the atmosphere, oceans, and cryosphere, and their interactions with the solid earth.
3. Characterizing the interactions of living organisms with the physical environment.
4. Understanding and monitoring the interaction of human activities with the natural environment.

The first of these themes is aimed at determining the composition, structure, and dynamics of Earth's interior and crust, and understanding the processes by which Earth evolved to its present state. Important properties of the mantle such as its composition, the spectrum of convective scales, and the relation between volcanism and tectonics are not understood. We will require measurements by seismic and other arrays of earth-based instruments, together with computer modeling and the monitoring of global gravity and magnetic fields, to fathom these processes.

The second theme is aimed at understanding the structure, dynamics, and chemistry of the oceans, the atmosphere, and the cryosphere. The interaction of these with the solid earth must then be detailed. Today we do not understand the factors that determine the global circulation of atmosphere and ocean, and the interaction of the atmosphere with surface geological and hydrological processes. The effects of biological processes on the hydrological cycle, climate dynamics, and geochemistry are major problems. We require satellite measurements, calibrated and validated from the ground, of these global-scale processes. For example, there is a pressing need for an instrument in orbit that can measure the rate of precipitation on the Earth—a major element in all models of the earth system.

The third theme deals with characterizing the interactions of living organisms among themselves and with the physical environment. This includes their effects on the composition, dynamics, and evolution of the ocean, atmosphere, and crust. The biosphere, for instance, controls the oxygen content and other aspects of the atmosphere, the oceans, and the solid earth. Yet land and ocean ecosystems are poorly understood or described today. Global measurements of biota from space, coupled closely with field experiments, are the key to better understanding in this realm. For example, ocean chlorophyll could be quantified by combining color measurements of the ocean with surface observations.

The fourth theme addresses human interaction with the natural environment. Human activity clearly affects the concentration of gases like carbon dioxide and methane in the atmosphere, as well as the amount of dust. Population increases and deforestation have uncertain implications for climate and genetic diversity. Conversely, many developments have made mankind more vulnerable to natural hazards. Some of these phenomena are best monitored from space, provided that proper calibration and validation are available.

RECOMMENDED PROGRAM: POST-1995

It is clear that to observe such an interactive and complex system as Earth we need both satellite and surface measurements. Satellites provide the global context for regional field studies, and most often are the only way to acquire global data. In particular, the steering group looks to a set of geostationary satellites to provide rapid synoptic images of the whole Earth. In addition, polar orbiters would provide high-resolution data and fill in the polar gaps. Special-purpose orbiters at various inclinations and altitudes would provide measurements as needed and communicate with instrumentation on the surface and in the atmosphere. A key requirement of these observations is their global completeness and simultaneity. Also, the observing system must be designed to assure continuous and consistent measurement over decades.

The volume of data collected by this many-faceted observing system will require faster, more automated, and more adaptable processing systems. Consistent formatting of different types of data from the atmosphere, oceans, and land will be essential. Better integration of modeling and observations will be another important aspect of future earth science systems. It is essential that data acquired over the globe be used both as inputs to these models and as tests for model predictions. This accomplished, scientists could use the entire Earth as a laboratory, following earth processes through their evolution. As always, advances in understanding require a mixture of empirical and fundamental approaches.

Specific recommendations given here, when implemented, will build on the results expected from the sensors and platforms of the NASA Earth Observing System (EOS), currently scheduled to fly as part of the Space Station complex in the mid-1990s. EOS, in turn, will build on its predecessor missions: the Upper Atmosphere Research Mission, the Navy's Remote Ocean Sensing System, the Ocean Topography Experiment, the Geopotential Research Mission, the Tropical Rainfall Mission, and the Magnetic Field Explorer. Other nations' missions, such as the European Space Agency's ERS-1 and Japan's Marine Observation Satellite-1, will also help define the specific parameters needed for adequate earth monitoring. EOS will be the next phase in the development of long-term measurement systems. But here the steering group looks beyond the initial deployment of EOS to lay out a series of specific recommendations for structure and programmatic content of a long-term mapping and monitoring system for Earth.

In this time period (1995 to 2015), the steering group suggests the following elements of an internationally sponsored program (U.S. responsibilities indicated):

1. *A Satellite-Based Observing System*

a. *A set of five geostationary satellites (two provided by the United States)* designed to carry a wide variety of instruments to cover the entire Earth for long-term measurements (replacement as required).

b. *A set of two to six polar-orbiting platforms (two to three provided by the United States)* to cover the polar areas above 60°

and to provide platforms for instruments that must be closer to Earth.

c. *A series of special missions that require other orbits.* Examples range from Shuttle-based instrument tests, to Explorer-type missions, to the Global Positioning System array of 18 satellites. With growing international interest in remote sensing of the Earth, the steering group expects an increasing proportion of joint or non-U.S. missions.

2. *A Complementary Earth-Based Observing System*

The steering group recommends the continuing development and deployment of *a system of earth-based measuring devices to provide complementary data to the space-based observing network.* The data from the network should be transmitted in real time and integrated with observations from space. This earth-based system is an essential element of any observing system for Earth; it measures effects that cannot be detected through remote sensing from space, providing increased resolution in regional studies, as well as calibrating and validating space observations.

3. *Theoretical Modeling*

State-of-the-art computing technology must be utilized for data analysis and theoretical modeling of earth processes. Modeling earth systems will require the best data sets possible, the fastest computers, and imaginative ideas from research. In turn, modeling can set the context and give direction to future observations.

4. *Data Systems*

A coordinated system for both archiving and disseminating earth-related data must be established. This is a call not for a central archive, but for a central authority or data management unit. This authority would establish formats and other conventions, identify data location, and provide easy access to all data as required. The data rates from the earth-observing system will be high, on the order of 10^{14} to 10^{15} bits per day. This will require much selective averaging and heavy use of new data storage and retrieval technologies. Automation of some phases of the selection and averaging process will be required.

THE ROLE OF NASA IN EARTH SCIENCES

The National Aeronautics and Space Administration is to be commended for the strong role it has played to date in earth sciences. Its efforts have ranged from studies of atmospheric, oceanic, and land surface processes to studies in the field of the solid earth sciences. For a long time, satellites have been used not only to sense properties of the atmosphere, ocean, and land surface, but also to define more precisely the shape of the Earth and to investigate the distribution of mass in its interior. As the new Earth Observing System (EOS) is developed, NASA should continue to play this key role in the development not only of space-based technology, but of the necessary earth-based systems and data systems as well.

The steering group endorses the position of the Earth Observing System Science and Mission Working Group that in future NASA missions "satellite-obtained data must be used in concert with data from more conventional techniques." The steering group agrees that, in addressing multidisciplinary problems, "observational capabilities must be employed which range in scale from detailed earth-based and laboratory measurements to the global perspective offered by satellite remote sensing." Clearly, such studies must be carried out together with the other agencies that support basic research in earth sciences, notably NSF, USGS, and NOAA, as discussed below. But a strong program within NASA itself must be maintained.

In particular, the steering group notes the importance of a strong program in the solid earth sciences. NASA could play a major role in a comprehensive program that deals with all of the most exciting and important questions in that discipline today. These questions include the origin of magmas, the driving forces for plate tectonics, and the generation of Earth's atmosphere. Moreover, high-resolution mapping of Earth's gravity field is essential if ocean surface topography measurements are to reach their full potential for ocean circulation studies. NASA's engineering capability in state-of-the-art technology (e.g., advanced satellite systems and data base management) is essential to the accomplishment of these objectives.

NATIONAL COORDINATION

Communication among the heads of the Office of Science and Technology Policy, the Office of Management and Budget, and the federal agencies involved in the civilian earth science effort is needed to develop coordinated programs and budgets. This requires full cooperation among the agencies involved: NASA, NSF,

NOAA, USGS, DOE, and others. The roles of these agencies relative to one another—NASA as a research and development agency, NSF as a supporter of basic research, and NOAA and USGS as operational, mission-oriented agencies in earth sciences—provide a test case for such cooperation. The steering group recognizes the importance of establishing clear roles, especially as researchers look to measurements on longer and longer time scales.

Coordination with the commercial sector is also essential. Plans are under way to operate the Landsat sensor package commercially, and the French are already flying a similar set of instrumentation on their Systeme Probatoire d'Observation de la Terre (SPOT) satellite series. The data are available commercially. Opportunities to fly other sensor packages, such as meteorological sensors, on leased spacecraft may occur in the future. Thus, any comprehensive program must include the commercial sector as a major player.

CONCLUSIONS

We now have the technology and the incentive to mount a "Mission to Planet Earth." The United States should implement this integrated program of fundamental research on the origin, evolution, and nature of our planet, its place in our solar system, and its interaction with mankind. The mission's feasibility has been demonstrated. We now need to act.

In order to mount this mission we need to deploy a major observational system with arrays of satellites and earth-based instrumentation for long-term measurements. In addition, we must bring into play new supercomputers, establish comprehensive data systems, and fund scientists, engineers, and other participants who make the program possible. This broad program will require support from many federal agencies, private industry, and the international community. NASA will play a key role in the implementation of the program.

3
Planetary and Lunar Exploration

BACKGROUND

Our solar system consists of nine known planets orbiting the Sun, and a large number of other objects: moons, asteroids, planetary rings, and comets. Among the mysteries that have preoccupied human thought throughout history are the mechanisms by which the solar system came into existence, the laws and physical processes that shape the evolution and behavior of planets, and the relationship of the solar system to the wider cosmos. The same questions continue to preoccupy modern planetary science as well.

Planetary studies illuminate some of the deepest and longest-standing scientific questions. Moreover, from a human perspective, planetary studies have additional significance. Planets are likely to be the only bodies in the universe capable of supporting advanced life. Among its other objectives, planetary science seeks to understand the formation of life-supporting planets and the conditions under which life arises and develops. The answers to these questions will shape our perceptions about our origins and our situation in the universe.

GOALS OF PLANETARY EXPLORATION

The scientific goals motivating planetary exploration are:

- To understand how the solar system originated;
- To understand how the planets evolved, including Earth and the planetary satellites, and to understand their present states;
- To learn what conditions led to the origin of life;
- To learn how physical laws work in large systems.

Each of these goals is explored below.

To understand how the solar system originated. Research aimed at understanding the origin of the solar system focuses largely on those objects thought to retain clues about the primordial conditions and processes that attended the system's formation. The most detailed clues come from investigations of comets, asteroids, and meteorites—small primitive objects that have changed little since their formation in the protoplanetary nebula.

The cold, volatile-rich matter of comets is thought to contain the most faithfully preserved samples of condensed protoplanetary material remaining in the solar system. The asteroids form an ordered assemblage of protoplanetary fragments that seem to remain near the locations of their original formation. They are thought to reflect the radial variation of conditions in the protoplanetary nebula. Laboratory analyses of meteorite fragments of asteroids and comets show the importance of the information these objects can provide. Detailed study of comets and asteroids is expected to fundamentally advance our understanding of the solar system's formation.

Planetary systems are believed to occur commonly in the universe as a result of the same processes that formed our own solar system. Failure to find such systems would force a fundamental revision of our theories about the origin of this planetary system and about star formation. Studies of star-forming regions and the discovery and study of other planetary systems will likely precipitate important advances in our understanding of the formation of the solar system, and in our understanding of planetary systems as a class.

To understand how the planets evolved. Because we live on Earth, a terrestrial planet, the evolution and environment of terrestrial planets is of special interest. Substantial advances in understanding can be realized by investigating, as a class, the terrestrial planets Mercury, Venus, Earth, and Mars, and other close analogs. In addition, studies of many of the outer-planet satellites and of the largest asteroids should reveal important information about solid planet evolution. Much of what we know of the terrestrial planets derives from ideas and concepts that originated in studies of Earth. Conversely, planetary investigations of objects that evolved under conditions far different from those on Earth may prod us to seek a deeper grasp of natural terrestrial phenomena, as well as a more complete understanding of Earth's history. By exposing circumstances in which concepts based on terrestrial analogs fail, planetary investigations help us define the limits of applicability of these Earth-based ideas.

As the world population increases and stresses the ability of our environment to accommodate it, terrestrial scientists will be called upon to model environmental impacts and to help develop tradeoffs between urgent resource needs and the consequences of meeting those needs. Models that can predict the properties of the widely varying atmospheres of the terrestrial planets will make this job much easier and enhance the credibility of scientists' pronouncements about this planet.

To learn what conditions led to the origin of life. Earth remains the only realm in which we know life has arisen. Our search to understand the origin of life involves several planetary questions: what are the physical conditions under which life arose, and have living organisms, either incipient or well-developed, arisen in other places where they can be studied? Presumably, living organisms arose out of an organic, prebiotic medium and were preceded by an interval of chemical evolution, which led more or less continuously into biological evolution. By understanding the formation of the planets, we will come to know the circumstances under which life arose on Earth. Many objects in the solar system seem not to have undergone substantial evolution since their formation. Some—Saturn's moon Titan, for example—probably carry important clues about the early material in which life arose. These and other objects in the solar system—including Mars—may have had prebiotic chemical species or harbored forms of incipient life, leaving evidence we can still collect.

Investigations of the composition of cosmic matter and primitive solar system matter show that the basic building blocks of terrestrial life, including amino acids, occur naturally, at least in trace amounts. One of the greatest challenges in understanding the origin and distribution of life is to determine just how widespread biological evolution may be in the cosmos. An important aspect of this question is the degree to which special terrestrial conditions were involved in prebiotic chemical evolution. Detailed chemical assays of comets, asteroids, and other primitive objects will reveal the extent to which life could have arisen directly from preplanetary matter without an interval of special processing to condition the chemical mix. This will provide important clues as to the possible ubiquity of biological evolution. (For a discussion on the origin of life from an exobiology perspective, see Chapter 7.)

To learn how physical laws work in large systems. Various phenomena are the unique result of the large scale of natural systems or arise from the very long times over which slow processes work. Investigation of large-scale physical processes involves virtually all of the objects in the solar system. The giant planets provide clues about properties of matter under high pressures; planetary interiors and magnetospheres demonstrate the curious behaviors of magnetized fluids and plasmas; and planetary atmospheres and surfaces present puzzles about the long-term evolution of complex interacting systems that constitute planetary environments and interiors.

Because these phenomena do not occur under normal laboratory conditions, it is only through direct observations in the solar system that we can understand them. Our current theories of planetary tectonism and cosmic plasma processes, for example, have developed in this manner. Since there is little prospect that in the foreseeable future we will make in situ measurements in other planetary systems, detailed investigations within our own solar system will continue to be the foundation upon which we build much of our understanding of natural phenomena throughout the universe.

ACHIEVEMENTS OF PLANETARY EXPLORATION

In the past 20 years or so most of the planets have been visited and several have been explored in some detail. A few highlights will be mentioned here.

The same physical processes operate on all the planets, but different starting conditions have led to a remarkable diversity of present states. This diversity is also influenced by violent collisions. The first recognized effect was cratering, prominent on the Moon, Mercury, and parts of Mars, where the largest basins and craters represent effects produced until accretion ended about 3.7 billion years ago. Craters are prominent on many of the moons of the outer planets, and a few of these moons seem to have been shattered by even larger impacts and then reformed. On Earth, geological processes have obliterated all but a few relatively recent craters, including one from an impact that may have caused mass extinctions at the end of the Cretaceous era some 65 million years ago. Recent theories explain the anomalous densities of Moon (low) and Mercury (high) in terms of enormous collisions with molten protoplanets in which the iron had already sunk to the center.

Great climatic change has been inferred for Mars, where abundant water once flowed, and for Venus, which may once have had the equivalent of a terrestrial ocean. Abundances of noble gases are remarkably different on Venus, Earth, and Mars, and different again on the parent bodies of meteorites. Ring systems are now known around all four of the giant planets, Jupiter, Saturn, Uranus, and Neptune; each ring system is totally different, and evidence is accumulating that some, at least, are transient. Jupiter's moon Io is the seat of many simultaneous volcanoes that shoot sulfur dioxide far above the surface. The sulfur and oxygen reappear as ions in a plasma torus enveloping Io's orbit. There they emit enormous amounts of ultraviolet radiation. More energetic ions populate the entire jovian magnetosphere and dominate much of its behavior. Saturn's large moon Titan is a terrestrial planet in many ways, but made of materials characteristic of the outer solar system. The dense nitrogen atmosphere contains methane clouds and a dark organic haze. A global ocean of liquid ethane is predicted. This variety of organic matter gives an environment analogous to what may have existed on a prebiotic Earth.

There is little doubt that more surprises and new concepts are still awaiting us and that the future of solar system exploration will be as rich as its past.

FUTURE PLANETARY EXPLORATION—A BALANCED PLANETARY PROGRAM

Progress toward realizing these goals requires a balanced program of basic science and exploration that includes studies of all planets, the Moon, and select asteroids and comets in our solar system. Concurrently, astronomical observations of star-forming regions and other planetary systems should be made. It is not safe to exclude parts of the system from study; experience has taught us to expect the unexpected. The initial exploration of all the planets, except Pluto, will be complete when Voyager flies by Neptune in August 1989. The next step is a more detailed examination of the planets to dissect the processes at work there.

So far, we have intensively studied only the Moon, Venus, Mars, and Jupiter. Beyond 1995, planetary exploration will shift increasingly toward orbiters, atmospheric probes, landers, sample returns, and perhaps manned exploration—the type of research required for a more complete understanding of the solar system. To complement these in situ investigations we will require laboratory experimentation and theoretical analysis as well. Data from spacecraft are largely responsible for the rapid advance in our view of the solar system. The steering group envisions that spacecraft investigations will continue to play this primary role.

Prospective pre-1995 Missions

Several projects that are now ready for launch or under development will set the stage for the vigorous solar system science and exploration program in the early years of the twenty-first century. Magellan will carry a radar system to map almost all the surface structure of Venus at a resolution of 300 m. Resolution of this quality will provide information key to comprehending the variety of evolutionary histories and processes undergone by the terrestrial planets. The Mars Observer mission will carry out the first global geochemical analysis of the martian surface and will investigate some properties of the planet's atmosphere. The Lunar Geoscience Orbiter will carry out a similar survey of the Moon.

The Galileo probe will carry instruments deep into Jupiter's atmosphere to measure its composition and physical structure.

The orbiter will also perform synoptic observations of the atmospheric dynamics, conduct a detailed investigation of the magnetosphere, and obtain detailed images and spectra of numerous jovian satellites. The Comet Rendezvous Asteroid Flyby will conduct detailed in situ investigations of a comet nucleus and will probe the physics of the comet-solar wind interaction. The detailed comet nucleus measurement will, hopefully, provide information about the conditions under which protoplanetary matter accumulated in the solar nebula.

The Soviet Union is expected to carry out major investigations of Mars and its satellite Phobos. This ambitious project is to include a new generation of analytical instruments for analysis of the surface composition of Phobos. In addition, the USSR is expected to carry out an asteroid rendezvous mission, involving a flyby of either Mars or Venus (probably the former). The Soviets are also considering a lunar polar orbiter mission, which would investigate the Moon's global chemical and mineral composition, magnetic fields, and temperatures.

Recommended Program: Post-1995

Over the 20-year interval from 1995 to 2015, the recommended program encompasses investigations of all of the major planetary bodies in the solar system along with selected satellites and primitive objects.

Terrestrial Planets

Landers, rovers, selected sample returns, and networks of automated observation stations on planetary surfaces will be the primary systems used to study the terrestrial planets. Specialized surface landers and rovers would allow the exploration of varied terrains and the surface material analyses that are necessary to ascertain the evolutionary histories of the planets. Analysis of selected samples returned to Earth will help us to determine both the character and the absolute dates of many of the major evolutionary events on the terrestrial planets.

Recommended Missions

1. Mercury:
    a. Orbiter
    b. Surface Landers/Sensor Network
2. Venus:
    a. Atmospheric Probe
    b. Surface Landers/Sensor Network
    c. Sample Return
3. Moon:
    a. Surface Landers/Sensor Network
    b. Scientific Rover
    c. Sample Return
4. Mars:
    a. Surface Landers/Sensor Network
    b. Scientific Rovers
    c. Sample Return

Outer Planets and Satellites

Each of the major outer planets—Jupiter, Saturn, Uranus, and Neptune—presents a complex, ordered system including the planets themselves, a magnetosphere, and a family of satellites and rings. Studies of the outer planets should include orbiting spacecraft to investigate all of these aspects. Atmospheric entry probes will reveal information critical to determining the composition and evolution of those planets. In situ studies of selected satellites will collect information pertaining to the primordial state of the volatile and organic matter in the solar system, and may yield clues about prebiotic chemical evolution.

Recommended Missions

1. Jupiter:
    a. Magnetospheric Polar Orbiter
    b. Deep Atmospheric Probe
    c. Io Lander
2. Saturn:
    a. Orbiter and Atmospheric Probe
    b. Deep Atmospheric Probe
    c. Ring Rendezvous

    d. Titan Orbiter and Probe
    e. Titan Lander (or floater)
3. Uranus:
    a. Orbiter and Atmospheric Probe
    b. Deep Atmospheric Probe
4. Neptune:
    a. Orbiter and Atmospheric Probe
5. Pluto:
    a. Orbiter

Primitive Bodies

In situ studies of the primitive bodies began with the missions to Halley's Comet. We will need rendezvous missions to other comets and asteroids to select the objects and the instrumentation to be used in later detailed studies. Investigations following rendezvous of a selected set of asteroids will allow us to determine their compositions and structures, as well as the variation of these properties in the main asteroid belt. We will thus obtain insights into processes in the protoplanetary nebula and explore early evolutionary mechanisms. In situ studies of comets and small outer solar system objects will permit analysis of the most complete and best-preserved samples of primitive matter, yielding clues to the origin of the solar system and of life. Return of samples from selected primitive bodies will allow the in-depth laboratory analysis possible only on Earth to contribute to this effort.

Recommended Missions

1. Comets:
    a. Coma Sample Return
    b. Nucleus Rendezvous and Sample Return
2. Asteroids:
    a. Multiple Rendezvous
    b. Sample Returns

Other Planetary Systems

Discovering and studying other planetary systems require the use of advanced telescopes in space. Planets disturb the motion of their central stars, and the evidence of these disturbances can be found by measuring the positions and motions of stars. In addition, such measurements can give information about the masses and orbits of the surrounding planets—information that can provide critical tests of our ideas about the formation of planetary systems and stars. *The steering group recommends the development of specialized telescopes capable of detecting planets at least as small as Uranus and Neptune around a large number of nearby stars.* The technology on which these telescopes depend is now within reach for use in space; the use of such telescopes in association with the Space Station would permit an observing program sufficiently long to allow the search for and study of planetary systems around a large number of nearby stars. Once other planetary systems have been discovered, there will be strong incentive to develop more sensitive instruments for further studies.

Recommended Missions

1. Space Astrometric Telescope

A Mars Focus

A Mars-focused program is recommended in parallel with the general program outlined above. However, this Mars-based program is not a substitute for a broader, balanced program of planetary exploration. There is no reason to expect that studying one or two planets in depth will allow us to understand how the entire solar system originated and evolved.

Planets and their environments exhibit behavior that, for fundamental reasons, cannot be predicted from first principles. The complexity of planetary environments is such that a planet can, in principle, exist in a large variety of states with the same conditions imposed from outside. The possible presence of living organisms further extends the variety of states in which a planet can persist. Finally, accidents of evolution can affect the state of a planet profoundly. A major challenge of planetary science is to trace the evolution of terrestrial planets, and enumerate the possible varieties and causes of their diverse environments. Meeting this challenge will require comparative studies of the terrestrial planets, including detailed studies of the changes that individual planetary environments undergo.

Of particular interest in the comparison of terrestrial planets is the puzzle posed by the triad of planets with atmospheres: Venus,

Earth, and Mars. The differences in their present environments and in their styles of evolution seem large in comparison with the differences in their sizes, locations, and overall compositions. Solving this puzzle is important to us because the differences between these planets occur in those aspects of their environments key to sustaining life.

Spacecraft investigations of Mars during the past 15 years reveal that the planet has undergone perplexing changes throughout its history. Although its surface is now dry and cold, there is clear evidence of a sustained, abundant flow of water during times past. Such changes in the martian environment directly pertain to long-term concerns about the behavior of Earth's environment. The prior presence of water on Mars raises important questions about its early, if temporary, suitability for life.

Images returned to Earth by Mariner and Viking spacecraft reveal spectacular geographical formations and deep-cut relief. It is evident that detailed study of the martian surface will yield information about the character of the planet's early environments, their arrangement in time, and perhaps clues as to the influences that produced such marked environmental change.

Of all the planets beyond Earth, Mars is the one most accessible to detailed study. It is relatively easy to reach with scientific spacecraft, and the surface is the most conducive to sustained operation of scientific instruments on mobile platforms. Furthermore, Mars is the only planet outside of the Earth-Moon system that we can currently consider for manned exploration and settlement.

Recommended Missions

1. Network of Geophysical Stations
2. Rover for Geology and Geochemistry
3. Sample Returns
4. Possible Human Exploration (The issue of the role of humans is discussed in a separate section of this report.)

CONCLUSION

The recommendations put forward here, if implemented, will advance our understanding of the solar system on the broad front that is needed to progress toward answering some of mankind's long-standing questions about the cosmos. A recommended Mars focus within that broad-based program will further our understanding of the terrestrial planets, including Earth, and will address pressing questions about planetary environments and their stability. The recommended investigations will also provide the information needed for proper planning of later manned exploratory missions to the Moon and planets.

4
Solar System Space Physics

BACKGROUND

Solar system space physics is concerned primarily with the sources and behavior of ionized gas (plasma) in the solar system. Plasma is sometimes called the fourth state of matter. Although solids, liquids, and electrically neutral gases are more familiar in everyday life, ionized gas is the most abundant state of matter in the universe. Plasma processes are essential to the physics of the Sun, as well as many other phenomena of the solar system and beyond.

The goals of solar system space physics are to understand:

- The physics of the Sun: its extended ionized atmosphere (the interplanetary medium), the magnetospheres, ionospheres, and upper atmospheres of the Earth, other planets, and comets, and the propagation of cosmic rays in the interplanetary medium.
- The processes that link solar variations to terrestrial phenomena. Such processes reveal basic physical mechanisms and influence many circumstances of human endeavor.

Study of the Sun, our nearest star, provides a firm basis for understanding stellar processes and astrophysical plasmas. Sunlight in the visible and near-ultraviolet portions of the electromagnetic spectrum sustains nearly all life on Earth. However, there are many subtle and less well understood influences of the Sun's emissions in other portions of the electromagnetic spectrum and in the form of high-speed ionized gases (the solar wind) and energetic particles. The interactions of these radiations with the neutral atmospheres and plasma environments of Earth and other planets cause a variety of physical phenomena. Some of these phenomena affect human activities, but all of them are of great scientific interest.

The flight of sophisticated instruments, first by high-altitude rockets beginning in 1946, and more recently by earth satellites and interplanetary spacecraft, has revolutionized research in solar and space physics. Indeed, this field of research has been one of the most successful of all of the space sciences. Observations above Earth's atmosphere have led to the discovery of solar ultraviolet, x-ray, and gamma-ray emissions, all of which arise from nonthermal processes. These observations provide the first undistorted view of the small-scale plasma structures that control such nonthermal processes.

At altitudes far above the atmosphere, there is a region containing electrically conducting plasma and a huge population of high-energy particles trapped in Earth's external magnetic field. This region is called the magnetosphere. Recent investigations reveal that Jupiter, Saturn, and Uranus have magnetospheres whose dimensions are on the order of millions of kilometers. While the magnetospheres of the planets in the solar system exhibit certain similarities, each is distinctive in detail. In addition, there are plasma phenomena associated with each of the inner planets—Mercury, Venus, Earth, and Mars—and with comets. A common element in all of these physical systems is the "solar wind," the flowing solar plasma that permeates the solar system. The region around the Sun where this solar wind occurs is called the "heliosphere." The boundary between the heliosphere and the interstellar medium has not yet been observed, but is estimated to be at a distance of 50 to 100 AU from the Sun, beyond the orbits of all known planets. The observation of this boundary between the heliosphere and the local interstellar medium is one of the central objectives of contemporary space physics. Solar system space physics research encompasses all of the plasma physical phenomena within the heliosphere.

THE SUN, SOLAR PROCESSES, AND VARIABILITY

The discovery of sunspots by Galileo in 1610 led to recognition, in the eighteenth century, of the 11-year cycle of sunspot number and other indices of solar activity. We know now that such activities have time scales ranging from a few seconds to at least 22 years. Giant solar flare eruptions occur at frequent intervals and release large amounts of energy into interplanetary space. This energy includes radiation that spans almost the entire electromagnetic spectrum and energetic particles sufficiently intense to kill exposed living matter in space. Reliable forecasting of such activity is one of the challenges of solar physics. The large-scale structures observed on the Sun are thought to be the macroscopic manifestations of small-scale plasma processes associated with turbulent magnetic fields. The Sun's brilliance and large apparent size offer unique opportunities for observing its underlying physical processes in great detail. Even the shape of the solar globe is not constant. It oscillates with periodicities from milliseconds to hours. There is an intimate connection between variations in the visible surface of the Sun (the photosphere) and processes deep within its interior. Recent work in studying these oscillations is making it possible to use a new technique, called "helioseismology," to penetrate the opaque brilliance of the solar surface. In much the same way as geophysicists study seismic waves to learn about conditions within Earth, solar physicists are exploiting natural oscillations of the Sun to probe its interior.

THE SUN-EARTH SYSTEM, THE MAGNETOSPHERE, AND THE AURORA

The million-degree outer atmosphere of the Sun expands in all directions at supersonic speed (about 400 km/s) and envelops Earth and all known planets. Earth's magnetic field deflects this flow of hot plasma from impinging directly onto its atmosphere. In the process, the magnetic field is compressed on the dayside and extended on the nightside into a long plasma magnetic tail like the visible tail of a comet. The solar wind power impinging on the magnetosphere is some 50 trillion watts, about 100 times the electrical generating capacity of the United States. A fraction of this energy, in the form of hot plasma, eventually penetrates the magnetic field and circulates through the magnetospheric system.

Some of this energy (about 300 billion watts) impinges on Earth's polar caps, generating spectacular auroral displays. It has been only within the last 5 years that we have been able to obtain a view of the aurora on a global scale using an imager on the Dynamics Explorer satellite. Despite many efforts, very little is known yet about how the energy enters the magnetosphere and circulates through the system, or where it is stored. Also little-understood are the physical processes that heat the solar wind plasma to several times its original temperature, and the eventual delivery of this energy into the high-latitude atmosphere. We are just entering a period that will generate sufficient data to allow a detailed understanding of Earth's magnetosphere and its environment.

THE UPPER ATMOSPHERE

There is an outer gaseous envelope surrounding all the planets (except Mercury) and some planetary satellites in our solar system. Such atmospheres absorb and redistribute globally the variable components of the solar energy falling on them. Planets with strong magnetic fields have dynamic neutral gas-plasma interactions associated with auroral activity and the convection of plasma around the planet. These processes are particularly prominent on Earth and Jupiter and have a strong influence on the global structure of the upper atmospheres of these planets.

On Earth, the upper atmosphere/ionosphere acts as the intermediary between the plasma-dominated magnetosphere and the bulk of the neutral atmosphere below. The region is highly complex. Interacting dynamical, chemical, radiative, and electrical variations occur there that couple the magnetosphere and middle atmosphere. To understand how these coupled elements interact to produce the great variability characteristic of the system is one of the major problems in solar-planetary relations. For example, the three-dimensional circulation of the thermosphere changes during and following geomagnetic storms; yet the consequences of the change of circulation on the temperature, density, composition, and electric currents of the region are poorly understood. Energetic solar particles penetrate the middle atmosphere and produce chemical changes in radiatively important species such as ozone, but their global consequences are not fully appreciated. Deeper in the atmosphere, solar-induced variations in the flux of cosmic rays may produce variations in the electrical structure of the lower atmosphere, but the effects of these variations on Earth's global electric circuit are not understood.

MAGNETOSPHERES OF OTHER PLANETS AND COMETS

Several planets possess magnetospheres because they have planetary magnetic fields; other planets, such as Venus, interact with the solar wind and form a downstream cavity, primarily through an interaction with the planet's ionized atmosphere. The largest known magnetosphere in the solar system is that of the planet Jupiter; its dimension on the sunward side exceeds 4 million km—about 6 times the radius of the Sun. The magnetic tail of Jupiter extends to at least the orbit of Saturn, a distance of 700 million km. Jupiter's magnetosphere is unique in that its plasma consists principally of oxygen and sulfur ions that are the ionized effluents of volcanic eruptions on the satellite Io.

Comets also have rudimentary magnetospheres. International Cometary Explorer (ICE; formerly ISEE-3) demonstrated this during its passage through the coma of the comet Giacobini-Zinner at a distance of 7000 km on the antisolar side of the comet's nucleus. The influence of the comet in the interplanetary medium was observed as far as a million kilometers away from the nucleus. The interaction of the coma of the comet with the solar wind appears to be yet another example of plasma processes that can occur when the hot solar wind impinges on the cool atmosphere of a solar system object. The March 1986 encounters of the Soviet, European, and Japanese spacecraft with comet Halley yielded further important advances in understanding cometary physics.

CONNECTIONS OF SOLAR SYSTEM SPACE PHYSICS TO LABORATORY AND ASTROPHYSICAL PLASMAS

Nearly 30 years of space research have clearly shown that many of the physical processes observed in the Sun, in the solar wind, and at the Earth occur throughout the universe. Detailed analysis and understanding of these, made possible by close-range and in situ observation, serve to shape our studies of more distant astrophysical phenomena. The rudimentary magnetosphere of Mercury is smaller than that of Earth by a factor of 20, while the size of the magnetosphere of a pulsar is believed to be comparable to the size of Earth's magnetosphere. By contrast, the magnetosphere of Jupiter is about 100 times larger than the magnetosphere of Earth, with an angular diameter about that of the Moon, even though Jupiter is 2000 times as far away. Even the size of Jupiter's magnetosphere, however, pales in comparison to that thought to surround the radio galaxy NGC 1265—approximately 100 billion km.

Physical processes that take place in Earth's magnetosphere have also been observed in other plasma "laboratories," ranging in size from Tokamak fusion devices (approximately 5 m) to solar flare kernels (less than 1000 km), to entire galaxies. Since the 1950s, plasma physics has developed along two separate, yet intimately connected paths: space and astronomical studies, and thermonuclear fusion research in terrestrial laboratories. Both are necessary for a complete understanding of solar processes.

NATURE OF THE FIELD

A substantial advance in the field of solar and space physics will require a major effort in each of its subdisciplines: solar and heliospheric physics, magnetospheric physics, cosmic-ray physics, and upper atmospheric and ionospheric physics.

Passive observational techniques are crucial to this field. They utilize instruments transported into space in a variety of ways: on high-altitude balloons and rockets; on long-lived earth satellites; and on interplanetary and planetary spacecraft, including planetary orbiters and entry probes. Active plasma experiments in the ionosphere and near-Earth environment have also played a significant role. Such experiments utilize electron and ion guns (small particle accelerators) and the controlled release of bursts of gas. In addition, ground-based observations play a continuing and important role. Basic theoretical work and modeling calculations underlie the entire field, guiding observational work. A balanced program including all of these elements is necessary for progress.

CURRENT FLIGHT PROJECTS

A first-order priority during the next several years is to maintain the operation of several existing satellites and spacecraft.

These include the Dynamics Explorer 1 (DE-1), the International Sun-Earth Explorers 1 and 2 (ISEE-1, 2), the International Cometary Explorer (ICE), the Active Magnetospheric Particle Tracer Experiment (AMPTE), the Solar Mesosphere Explorer (SME), the Pioneer Venus Orbiter (PVO), and the four outer-planets-heliosphere spacecraft Pioneer 10, Pioneer 11, Voyager 1, and Voyager 2. All of these spacecraft are devoted wholly or partially to solar and space physics and are continuing to provide valuable data. They have special importance because of the limited prospect for new missions during the next several years. The August 1989 encounter of Voyager 2 with Neptune is an event of particular interest.

PROSPECTIVE PRE-1995 MISSIONS

The Galileo orbiter/probe to Jupiter, the solar polar orbiter (Ulysses), the Upper Atmosphere Research Satellite (UARS), and the International Solar Terrestrial Physics Program (ISTP) are the most important new elements of the solar and space physics program from now until 1995. In addition, plasma and plasma wave instruments, and an electron gun on a Shuttle or free-flying mission in Earth's ionosphere are planned for this period. Also, the Mars Observer (MO) spacecraft will carry a magnetometer for improved study of the magnetic field of Mars. Galileo and Ulysses are ready for flight, but their launching dates have been long postponed and are still uncertain. UARS, ISTP, and MO launches will probably take place in the early 1990s.

A substantial portion of the Galileo instrumentation is intended for second-generation study of Jupiter's magnetosphere. The Ulysses mission represents a pioneering effort to observe the interplanetary medium, solar energetic particles, cosmic rays, and the Sun's atmosphere at high solar latitudes—all classical phenomena in interplanetary and solar physics. The goal of the UARS program is to understand processes that control the structure of Earth's stratosphere and lower thermosphere. This includes the ozone layer and its response to natural and artificial perturbations. ISTP is being developed jointly by the United States, Japan, and the European Space Agency. The main objective is to develop a comprehensive global understanding of the generation and flow of energy from the Sun through the interplanetary medium and into Earth's space environment. In addition, ISTP seeks to define the relationships between the physical processes that link different regions of this dynamic system. Finally, the High Resolution Solar Observatory (HRSO), with a spatial resolution of about 70 km, will provide initial data on plasma-magnetic field interactions related to the solar dynamo and energy transport in the solar atmosphere.

RECOMMENDED PROGRAM: POST-1995

The steering group has identified a number of forward-looking programs that will address basic scientific problems in solar system space physics. Some of these have been the subjects of prior spacecraft and mission design studies.

1. *Ultraviolet and X-Ray Telescopes of the Highest Practical Resolution.* Such instruments are required to investigate small-scale (1 to 100 km) processes associated with transient and turbulent regions. The evolution and dynamics of such small structures are believed to be essential to the understanding of large-scale structures such as active coronal regions, flares, and the origin of the solar wind.

2. *Solar Probe.* This is an exploratory mission, contemplated for the mid-1990s, to investigate the nearby neighborhood (altitude of approximately 1.9 million km) of the Sun. The basic scientific goals are to explore the solar atmosphere, which is now known to us only through remote images. It will also investigate the sources of the solar wind. The great technological challenge here will rest with the development of a heat shield capable of withstanding the high temperatures encountered as the spacecraft approaches the Sun.

3. *Other Advanced Spacecraft.* Some of these are in the conceptual phase: viz. the Earth Observing System (EOS), the Solar Terrestrial Observatory (STO), and the Advanced Solar Observatory (ASO).

4. *Advanced Missions to the Outer Planets.* Projects in this category that have been the subject of specific workshops are: the Cassini Mission, involving a Saturn orbiter for detailed, synoptic studies of Saturn's magnetosphere and other planetary purposes, and a Jupiter Polar Orbiter (JPO) as a successor to Galileo.

5. *Interstellar Probe (IP).* In this project, planned for the turn of the century, a spacecraft would escape the solar system at great

speed (about 80 km/s) and enter the local interstellar medium within 10 years. Such a spacecraft, if launched in the year 2000, would overtake the Pioneer and Voyager spacecraft (launched in 1972, 1973, and 1977) within 5 years and proceed ahead of them into the outer solar system.

6. *Remote Sensing of Magnetospheric Plasma.* This is illustrated by the comprehensive imaging of aurorae on a global scale by DE-1. This technique, emphasizing selected energy bands in the ultraviolet and x-ray regions of the spectrum, has great future potential for remote sensing from locations at the Lagrangian points, on the Moon, and on high polar-orbiting satellites. An important new technique for remote sensing of magnetospheric processes by observing escaping neutral atoms has also been demonstrated recently and should be fully developed. Such investigations will advance our understanding of the entry and circulation of plasma within Earth's space environment—something addressed prior to this time principally by local measurements, often separated by long time intervals. In some sense, the magnetospheric problem is the inverse of that of the Sun, for which we have global measurements but lack details about the local environment.

7. *Active Experimentation.* This is a valuable technique in the study of space plasmas. It involves the injection of gases, electromagnetic waves, or particle beams into the natural environment and the observation of the resulting interactions. Some of the most significant results have come recently from the creation of artificial comets by the Active Magnetospheric Particle Tracer Explorer (AMPTE), a collaborative program between the United States, Germany, and Britain. Further experiments of this type, as well as observation of the solar wind interaction with natural comets, are planned. Studies of dust-plasma interactions using active experiments may be valuable in helping to understand the formation of the solar system. In addition, it may be possible to perform basic plasma physics experiments in space, including studies of plasma confinement in fusion systems, without the presence of walls.

CONCLUSIONS

The goals of solar system space physics address not only the basic physics of magnetized plasmas in the solar system, but also the complex energy transfer beginning at the Sun and propagating through the interplanetary medium to the magnetospheres, ionospheres, and upper atmospheres of Earth and other planets.

Because of its proximity, the Sun is the only star whose interior structure and atmosphere we can study at high resolution, thereby providing information about physical processes important to all stars. Magnetospheres, the magnetized plasma atmospheres of Earth and the planets, are now known to exist throughout the universe—around pulsars, radio galaxies, and accreting stars. The study of plasma processes that regulate the structure and dynamics of planetary magnetospheres has contributed significantly to the development of basic plasma physics. Further, there are important connections between solar and space physics and earth system studies. Variations in the solar output of radiation and charged particles have substantial effects on the magnetosphere, ionosphere, and upper atmosphere. Even small variations of the solar luminosity may affect Earth's weather and climate. Stratospheric and mesospheric ozone responses to incoming charged particles from solar flares are examples of such processes.

Finally, the understanding of the space environment near Earth has direct practical aspects, aside from its research value. Space is being used increasingly for scientific, commercial, and national security purposes. Space vehicles must function continuously in the near-Earth environment, subject to the influences of the Sun, the magnetosphere and upper atmosphere, and cosmic radiation. In addition, these elements of the space environment will become particularly important to humans should they attempt to spend long periods of time in space. This is true not just of manned missions within the magnetosphere and the interplanetary medium, but also on the Moon and possibly on Mars. An improved understanding of solar variability and the perils of solar flare radiation is mandatory. Substantial advances in our ability to operate space-based systems safely and reliably will result from the basic studies outlined in this chapter. (For a general statement on the presence of human activity in space, see Chapter 9.)

5
Astronomy and Astrophysics

BACKGROUND

The universe we perceive today appears much more complex in its design and more mysterious in its ways than anyone could have predicted in generations past. Early in this century, the stars were thought to form an unchanging cosmic tapestry, remote and inaccessible. But within our lifetime modern technology and physical theory have let us glimpse a far grander cosmological scheme. Our Milky Way galaxy is one of a myriad of island universes, flying apart after an initial "big bang" that not only determined the structure of the universe, but seems to have determined the physical laws that govern the behavior of elementary particles. Today, astronomers address questions that would have been framed in earlier times only by philosophers.

A major contribution to our expanded world view has come from new generations of astronomical instruments. Larger optical telescopes collect more photons, and electronic detectors and advanced spectrographs yield more detailed information about the physical state of matter throughout the visible universe than their predecessors. Observations using wavelengths outside the optical window began with radio astronomy and its revelation of entirely new aspects of the universe. A dramatic improvement came with the advent of space vehicles that, by carrying detectors above the atmosphere, allowed observations throughout the range of the electromagnetic spectrum—from x ray to infrared—that had previously been blocked by the Earth's atmosphere. High-energy astronomy using x-ray and gamma-ray observations has revealed violent phenomena invisible at other wavelengths. Ultraviolet astronomy allows stellar astronomers to study that region of the spectrum in which the spectral lines of key elements occur, while the development of infrared astronomy reveals yet another aspect of the universe: the interiors of the dark dust clouds where stars and planets are born.

The present rapid expansion in astronomy is not a transient phenomenon. There is no evidence that we are approaching a state of complete scientific knowledge; in contrast, a new age of space astronomy is just beginning—the era of "great observatories" in space. The period from 1995 to 2015 will be a pivotal one, since the time scale for planning observatories of the future is a decade or more. In this study, therefore, the steering group first examines the scientific considerations that drive the program. It then sets out the expected state of space astronomy 10 years hence and projects the classes of instruments that will be necessary during the succeeding 20 years. The program is bold but realistic.

Three basic principles guided the steering group's planning: (1) astronomy requires access to the entire electromagnetic spectrum, an access available only through space techniques; (2) the ability to obtain higher angular resolution will result in powerful new insights into stars, planets, and galactic nuclei; (3) telescopes with greater collecting area, higher resolution, and more efficient spectrographs will be needed in every wavelength band to observe the farthest and faintest objects. These basic thrusts provide the framework and the focus of the proposed program.

MAJOR SCIENTIFIC QUESTIONS

The astronomy and astrophysics program is designed to answer a set of fundamental questions that deal with three general topics:

1. *The early universe, including the large-scale structure of the universe, dark matter, and the formation of galaxies.*

2. *Strong force regimes: physics of gravitational collapse and attendant active processes.*

3. *The formation of stars, planetary systems, and the origin of life.*

Interest in these questions has persisted over the last two decades, and they remain as valid guides for the next several decades.

The questions related to the events of the early universe—starting with the basic puzzle, "How did the universe begin?"—appear to have a surprising connection with current theories of the fundamental forces between elementary particles. Various versions of current "Grand Unified Theories" of the fundamental physical forces lead to evolutionary models of the universe that require the average particle energy to be 10^{15} GeV (an equivalent temperature of 10^{28} K) only 10^{-35} s after the creation event (accelerators of the sort that physicists now use would need to be one light year long to produce particles of that energy). Before that instant of time, theories assert that the universe was composed of rapidly expanding matter in a primeval state; the tiny volume of the universe suddenly cooled down and was thus transformed into a hot gas in a manner somewhat analogous to supercooled water expanding when it freezes. After a brief period of extremely rapid inflation, the rate of expansion then settled down to the level that is deduced from present observations.

Such theories, resulting in a cosmology dubbed "inflationary" due to the initial period of rapid expansion, require a geometrically flat universe in which expansion slows down forever, growing to larger and larger dimensions but never halting. The amount of matter in such a universe must be about 100 times that which is deduced from all the visibly luminous material in all the stars of all the galaxies, and about 10 times the amount of "ordinary" matter, comprised of the familiar protons, neutrons, and electrons, believed to be present, although largely invisible. A number of independent arguments support the idea that a large fraction of the matter in the universe is "dark matter," whose nature is still a subject of speculation. Some "ordinary" dark matter could be very faint stars (brown dwarfs). But the additional dark matter required by inflationary theories cannot be ordinary matter. Rather, it must exist in some exotic form, such as massive neutrinos, or conjectured particles such as axions, photinos, or gravitinos. None of these particles has yet been observed. At an even more speculative level, the dark matter could be in the form of massive black holes or cosmic strings—infinitesimally thin (about 10^{-30} cm) and enormously massive (about 10^{22} g/cm)—stretching across the entire universe. Produced abundantly in the "big bang," cosmic strings would not be directly visible, but might be detectable by their large gravitational lensing effects.

In the past year, an even more exotic possibility has arisen with the "superstring" theory of matter. Highly conjectural, this theory has the attraction of being specific, since it leads to the concept of an 11-dimensional space-time with well-specified internal symmetry. The most straightforward argument leads to two symmetry groups, one of which gives rise to the universe of particles with which we are familiar. The other symmetry group would generate a completely different set of particles that we can detect only through their gravitational interaction. Thus, our universe might be coexistent with a second "shadow universe" of particles that interact through forces we can experience only through their gravitational effects. The methods of astronomy alone can measure these.

These explanations are far from established, yet the underlying fact is that the "dark matter" in the universe is present and remains to be understood. The subject of cosmology has always had close ties to theories of fundamental physics, and this continues to be the case. Our knowledge of the universe on a cosmic scale is still limited, and the methods of space astronomy, using the instruments proposed here, will bring vital new knowledge and understanding.

The second major topic, the behavior of matter under extreme astrophysical conditions, also emphasizes the important relationship between astronomy and modern physics. We know that white dwarf stars and neutron stars exist, and that supernovae mark the end of the life of a star. The physical processes associated with these phenomena are far from understood, but are of the most fundamental interest. Neutron stars, for example, are in a sense the largest nuclei of all, and their behavior is determined by the forces that act when matter is as dense as that in the atomic nucleus. Such behavior is far from simple. The complex phenomena associated with pulsars—which are rotating neutron stars—demonstrate this.

The explosion of a supernova is also ill-understood, and is equally important. The heavy elements that make life possible are generated in these explosions, but theory is only beginning to show how the explosion occurs and proceeds. Another intriguing question is the physical state of the stellar remnant of a supernova—a black hole or a neutron star. There are x-ray sources in binary star systems in which an unseen companion body is so massive that theory implies it might well be a black hole. Verification, it appears, will come only through space astronomy.

On a still larger scale, the powerful energy machines in quasars and active galactic nuclei seem to require a black hole of a million to a billion solar masses at the core. The program outlined for the period from 1995 to 2015 will probe ever closer to the black hole (or other large concentration of mass) at the heart of these mighty engines. There the principles of physics will be tested to their limits, for strong gravitational fields such as those near black holes are the least understood and tested of the fundamental force fields. In the study of black holes, Einstein's theory of general relativity receives its most severe test.

The third major topic has a number of aspects that have developed only within the past few years. The Infrared Astronomical Explorer Satellite (IRAS) has sent back a treasure of surprising information relative to the formation of planets and stars; interferometric astronomy has been used to detect planets by the wobble they induce in their companion star. Finally, it appears that even planets as small as Earth might be detected by large telescopes and imaging interferometers. These can then study the characteristics of the planetary atmospheres. If life—particularly earth-like life—is present, we may find evidence for it in the molecular constituents of those planetary atmospheres.

THE EVOLUTION OF SPACE ASTRONOMY

The early years of the space age brought great surprises. Some of the developments came from ground-based discoveries that at first seemed to be unrelated to the subjects of space astronomy; later developments have shown a multitude of surprising cross-links. The picture now evolving of the unity of modern astrophysics explains the steering group's emphasis on the need to have simultaneous access to all regions of the electromagnetic spectrum.

Consider, for example, the events that followed the discoveries of the early 1960s. On the ground, radio and optical astronomers, working together, discovered quasars. The precise nature of quasars is still not understood, but they are the most powerful celestial engines that have been found in the universe, capable of radiating the power of thousands of Milky Ways from a volume that is only a trillionth that of an ordinary galaxy. It turned out that they are powerful emitters of x rays and gamma rays as well. The physics of their excitation may be closely related to the observed x-ray behavior of active binary stars, and it is clear that progress will come from a union of high-energy space astronomy with radio and optical observations.

The first evidence of cosmic x rays came in 1962. It was a complete surprise, unanticipated by any theories. As observations proceeded, it became clear that many stellar x-ray sources are in systems of binary stars, with the x rays being generated by matter from one member falling onto its companion star. The companion is generally a highly compressed star, sometimes a white dwarf or a neutron star—perhaps, in some cases, a black hole. These identifications were made possible when x-ray, optical, and radio astronomers joined forces.

Other surprises marked the early years of the space age. The discovery of the microwave background showed that the "big bang" concept of cosmology was fundamentally correct, and led to the construction of the Cosmic Background Explorer (COBE) mission, an Explorer-class satellite that will probe fundamental aspects of the relict radiation from the early universe. Pulsars were discovered by radio astronomy, and early rocket observations showed that the pulsars were emitting x-ray pulses as well. One of the earliest and most surprising observations made at gamma-ray wavelengths was that of gamma-ray bursts, detected by instruments aboard the Vela satellites, a series of satellites launched to monitor the nuclear test ban treaty of 1963.

As the era of telescopes in space began, the surge of discovery continued. The first x-ray astronomy satellite, Uhuru, generated a comprehensive catalog of x-ray stars, galaxies, and clusters of galaxies, and strong evidence was found for a black hole in the constellation Cygnus. In 1973, the Copernicus mission offered the opportunity for ultraviolet spectroscopy of galactic sources and the interstellar medium. Hot interstellar gas (about 500,000K) traced a lacy web along colliding fronts of expanding cosmic gas bubbles, the debris of supernova explosions. The first direct evidence for interstellar heavy hydrogen (deuterium) was also obtained in these pioneering observations. Very-long-baseline interferometry, using radio waves, revealed fine structure in the tight nuclei of quasars. Motions there were measured that appeared to be faster than the speed of light. The law of physics that prohibits this behavior for real motion is presumed valid, so the current belief is that this phenomenon is an optical illusion caused by bulk relativistic motions generated by the core of quasars. Furthermore, interferometry methods developed for radio astronomy appear to be directly applicable to high-resolution optical studies. The ambitious plans for infrared and optical interferometry during the period covered by this study will draw directly on this experience.

With the launch of COS-B in 1975, gamma-ray astronomy came into its own. Only 4 of the 26 high-energy gamma-ray sources discovered have been identified with known quasars and pulsars. The nature of the remaining sources, forming a catalog of UGOs (unidentified gamma-ray objects), is baffling.

Discoveries in all wavelength bands revealed the need for various space telescopes. The first of these, a powerful x-ray telescope, was orbited aboard the Einstein Observatory (HEAO-2) in 1978. At the limits of the universe, x-ray quasars were found to shine so powerfully that they were detected more readily than their optical counterparts. Close by, even the faint dwarf stars of the Milky Way were detectable x-ray sources, sometimes flaring to thousands of times the brightness of the largest solar flares. Early-type giant stars were found to be such prolific x-ray sources that often their entire surface seemed to be excited as though by a giant flare.

Early in the 1980s, the Infrared Astronomical Satellite (IRAS) opened a new wavelength band to investigation when it discovered a quarter of a million new infrared objects. As it focused on the young star Vega, 27 light years from Earth, IRAS detected what appears to be a protoplanetary system extending out to about 15 billion miles from the star. Thus opened a new era in planetary astronomy.

The IRAS infrared telescope was the latest space-age telescope to be launched. It represents, in a sense, the transition to a new era in space astronomy—the age of the great observatories. We now realize that we need long-lived telescopes in space at all wavelengths. In projecting the status of space astronomy in 1995, the start of the two-decade period addressed by this study, the general outlines of the program are clear.

The Hubble Space Telescope (HST), the first major observatory, is now awaiting launch. It will bring two major advances to astronomy by freeing a telescope from the limitations imposed by Earth's atmosphere. First, the telescope can observe far into the ultraviolet part of the spectrum, where many of the most important elements emit their fundamental spectral lines. Second, it will be free of the atmospheric blurring effect called "poor seeing," and will capture the finer details of celestial objects. At the same time, it will detect much fainter, more distant stars and galaxies, because the sharp images it can produce will stand out with greater contrast against diffuse sources of light in the night sky.

Gamma-ray astronomy exploits the highest energy range of the electromagnetic spectrum, a difficult band to study since the sources yield so few photons on Earth. The Gamma Ray Observatory (GRO) will cover the spectral band from about 1 to 1000 MeV, and represents perhaps the ultimate capable with present-generation instrumentation.

The Hubble Space Telescope and the Gamma Ray Observatory are due to be launched as soon as the consequences of the *Challenger* accident are resolved. The evolution of space astronomy in the years that immediately follow these launches has been set out in the NRC report *Astronomy and Astrophysics for the 1980s*, in which the Astronomy Survey Committee (ASC) formulated a program for the next decade. The steering group found in that report a reliable road map for the next 10 years of space astronomy. With the exception of the Large Deployable Reflector (LDR), which is one of the steering group's major recommendations for the period from 1995 to 2015, all the components cited by the ASC should be well advanced by 1995.

As the ASC suggests, the next major step in x-ray observatories will be the Advanced X-ray Astrophysics Facility (AXAF). The AXAF aperture will measure 1.2 m in diameter, twice that of the Einstein Observatory, and will contain a nest of seven reflectors. It will have 10 times the angular resolution, 50 times the sensitivity, twice the spectral range, and 1000 times the energy resolution of the Einstein Observatory. The kind of work conducted earlier by the Einstein Observatory will be greatly accelerated with AXAF.

The development of infrared astronomy may soon rival radio,

optical, and x-ray astronomy. The Space Infrared Telescope Facility (SIRTF), which will be 1000 times as sensitive as IRAS, will join the family of great observatories when it has been placed in orbit. It will bring millions of infrared sources within observing range.

The design technology for each of the great observatories to follow the Hubble Space Telescope is well in hand. This entire constellation of spacecraft can be in place before the mid-1990s. The ability to conduct coordinated observations at various wavelengths will be one of the great benefits of flying these instruments simultaneously.

The Astronomy Survey Committee recognized that there is an equally important, far less expensive, component of the space astronomy program that must not be neglected. This "exploratory" program consists of smaller, ad hoc projects that prepare the way for major thrusts of the future. Explorer-class missions exemplify this kind of project, and several of these will become operational during the coming decade. The X-Ray Transient Explorer (XTE) will allow in-depth investigation of the bursts, pulse, and other transient phenomena characteristic of active x-ray sources. The Far Ultraviolet Explorer (LYMAN) will allow observation of atomic and molecular spectra over a wide range of energies that are beyond the wavelength limit of the Hubble Space Telescope. The orbiting very-long-baseline interferometry radio telescope (QUASAT) will permit the expansion of interferometry to baselines larger than Earth, and will obtain resolution of quasars and other active objects that approach 1 μarcsec. In the spirit of the Explorer program, other interesting exploratory projects will surely arise over the coming decade, and NASA should stand ready to exploit these opportunities.

RECOMMENDED PROGRAM: POST-1995

The advance of space technology, lifting power, and space assembly capability offers great promise for a number of new ventures in space astronomy. In the first section of this chapter, it was shown that these can be classified, broadly, into two categories: instruments that will give the kind of breakthrough in angular resolution that will allow the study of fundamental phenomena, and telescopes of great collecting area and spectroscopic capability that will carry on the tradition exemplified by the large telescopes of Earth such as the 100-inch Mt. Wilson telescope and the 200-inch "glass giant" of Palomar.

The program for astronomy and astrophysics can be classified more explicitly as follows:

1. *Imaging Interferometry*
   (a) *Large Space Telescope Array*
   (b) *Long Baseline Optical Space Interferometer*
   (c) *An array of very-long-baseline interferometry (VLBI) stations in space*
2. *Large-Area and High-Throughput Telescopes*
   (a) *A large deployable reflector (LDR) for submillimeter studies*
   (b) *An 8- to 16-m optical space telescope*
   (c) *Large-area telescopes for the energy range 20 keV to 2 MeV*
   (d) *A large Compton telescope for spectroscopy, 0.1 to 10 MeV*
   (e) *Large gamma-ray telescopes for energies above 2 MeV*
3. *AstroMag, a massive cosmic-ray analyzer in space*

Some of these projects, which are explored more fully below, are logical successors to those now under way. On the other hand, some are novel and will require new research programs. Realization of the recommendations for interferometry, for example, will require a variety of preparatory technological studies during the coming decade to establish the background for major missions in this field.

Imaging Interferometer (Optical and Infrared)

A two-step plan for interferometric projects can be foreseen. The earliest mission would probably be a large array of telescopes. A reasonable projection would be for an array of several telescopes mounted on a structure 100 m or so in diameter. This structure could be tetrahedral, supporting nine 1.5-m telescopes, three along each leg of the base with a signal-processing cabin at the fourth vertex. This array could map a field of about 0.5 arcmin with a resolution of 0.5 milliarcsec at a wavelength of 5000 angstroms. It would have a sensitivity higher than that of the Hubble Space Telescope and would provide images 100 times sharper in angular detail.

The next step, to a long-baseline space interferometer, is more challenging and demands far more engineering study. It would consist of telescopes whose baselines would range from tens of meters to many kilometers. If station-keeping technology and the art of metrology can advance, these instruments might be independently orbiting, perhaps at the stable Lagrangian point. They might also be constructed on the Moon.

The VLBI radio array in space would be a logical extension of the QUASAT project. The project could be an evolutionary one, with standard radio telescopes placed in successively higher orbits. It might well be a cooperative international project.

The Large Deployable Reflector (LDR)

One of the highest priority missions for the United States astronomical community is the Large Deployable Reflector (LDR) being designed for work in the far-infrared and submillimeter regions. The desired aperture is in the 20- to 30-m range, and the instrument would need to be assembled by astronauts at a space station. It will provide high angular resolution (1 to 2 arcsec at 100 μm and a diffraction limit of 0.3 to 0.6 arcsec at 30 μm. Improving our present capabilities nearly 1000 times, the LDR would join the suite of great observatories. It would be a natural sequel to the SIRTF mission.

An 8- to 16-m Telescope for Ultraviolet, Optical, and Infrared Wavelengths

About the same time as we deploy the LDR, we will require a filled-aperture telescope of 8- to 16-m diameter, with ambient cooling to 100K for maximum infrared performance. Such an instrument will monitor the range of wavelengths from 912 angstroms to 30 m and will follow on 10 to 20 years of study with the HST and ground-based 8- to 10-m telescopes. It will complement the space interferometer and provide images 6 times sharper than the HST. It will also be far more sensitive than HST because of its large aperture and small image size (10^{-2} arcsec for a 15-m diameter telescope in visible light). In addition, it would surpass the HST in the study of distant quasars and the evolution of galaxies, including the formation of binary systems and planets. Finally, direct imaging of sources with a filled aperture will offer substantial advantages over images reconstructed with model-dependent techniques from interferometric data.

X-Ray Instruments of Large Area and Throughput

The Advanced X-Ray Astronomy Facility (AXAF) will elicit new scientific questions we cannot yet foresee. Nevertheless, we will probably need an instrument that can perform high time-resolution studies and high-resolution spectroscopy and make observations at higher x-ray energies. This might take the form of a Very High Throughput Facility (VHTF) that could perform the spectroscopy, or an x-ray timing facility that could look for rapidly varying events correlated with the results from gravity wave detectors. A High Energy Imaging Facility (HXIF) might allow the first in-depth exploration of the hard x-ray/soft gamma-ray region of the spectrum, roughly from 20 keV to 2 MeV. This instrument would address fundamental questions about anisotropy in the early universe. It would also give us insights into compact objects, stellar collapse, and star formation. At present, there is a gap in the electromagnetic spectrum in the 20-keV to 2-MeV range that must be filled. An instrument observing in this region must have a large area, since incoming photons are few.

Gamma-Ray Telescopes

We will have gained several years of experience with the Gamma-Ray Observatory (GRO) by 1995, and the results will surely influence plans for the gamma-ray observatories of the future. The gamma-ray domain is of vital interest for two reasons. First, the nuclear lines characteristic of supernovae and other high-energy phenomena appear there. Second, the character of very high energy gamma rays is quite unexpected. An Advanced Compton Telescope or other spectroscopic device can provide the capability to carry out gamma-ray spectroscopy effectively. Above 50 MeV, where basic pair production processes do not generate a limiting background, we should be able to realize an angular resolution approaching 1 arcmin.

Cosmic-Ray Research

Programs in particle astrophysics will explore new regions of the spectrum at greatly improved levels of sensitivity and resolution. Many of the current cosmic-ray problems should be accessible to ASTROMAG, a superconducting magnet spectrometer with capabilities comparable to those used by laboratory physicists at the large accelerators. The facility should be an early project in the era 1995 to 2015.

CROSS-LINKS WITH OTHER DISCIPLINES

The Sun is our closest star, and it provides, along with the solar plasma, a basic reference point for many astrophysical problems. Since the Sun is a star, solar astronomy and stellar studies are closely linked. The interferometric instruments projected for the period from 1995 to 2015 are likely to open a new era in stellar studies, as spots, flares, and other phenomena begin to be studied directly on other stars. The Einstein Observatory has already demonstrated that studies of stellar coronas can be carried out by x-ray instruments.

Studies of other planetary systems and, possibly, of life outside the solar system will be a goal of the various interferometric systems and for the 16-m optical telescope. The instruments, if properly designed, should be capable of detecting planets—even small planets like Earth—in orbit about nearby stars. They might be able to study the atmospheric constitution of these planets as well. If abundant atmospheric oxygen or other evidence suggestive of biological processes is found, there is potential for a new link both with planetary science and the life sciences.

CONCLUSIONS

The recommendations of the Astronomy Survey Committee form a valid basis for assessing the expected status of astronomy and astrophysics in 1995. The major science goals for astronomy can be formulated with some reliability. A period of great scientific productivity during the time from 1995 to 2015 can be expected. The guiding principle is to assure access to the entire electromagnetic spectrum, to obtain high (milliarcsec to μarcsec) angular resolution from radio to ultraviolet wavelengths, and to build telescopes with large collecting areas and spectrographs of high throughput. At the same time, a vigorous exploration program, carried out by Explorer-class satellites, promises to supply a sound basis for studies in the even more distant future.

6

Fundamental Physics and Chemistry: Relativistic Gravitation and Microgravity Science

OVERVIEW

A common link among traditional space sciences such as space astronomy and astrophysics, planetary exploration, and solar plasma physics is their use of spacecraft for their observations. One of the objectives of this study was to determine whether there is likewise a potential to use space vehicles as laboratories in which fundamental physical and chemical laws might be investigated. The answer is decidedly positive. Spacecraft can provide a unique environment for at least two kinds of studies: those that would further our knowledge of relativistic gravitation and those exploring fundamental processes that require very small gravitational forces or very small gravitational gradients. The implications of using space vehicles for the study of general relativity have been understood for some time, and a specific strategy for investigations of relativistic gravitation from spacecraft after 1995 has been set forth here. On the other hand, the implications of exploiting the nearly gravity-free environment of space to study basic properties of matter have not been well delineated before, and the identification of opportunities in this realm is an important new achievement—one of the most exciting to emerge from this study.

A.
RELATIVISTIC GRAVITATION

BACKGROUND

General relativity relates the geometry of space and time to the distribution of matter in the universe. Gravitation is the consequence of the way this space-time geometry affects the movement of matter in space. As a theory, general relativity is well developed; it has important consequences that can be tested. There are three classical tests of general relativity in weak fields—such as those near the Sun or Earth. The first involves the precession of the perihelion of a solar system object such as the planet Mercury. The second utilizes the deflection of light passing close to the Sun. The third involves the gravitational red shift of spectral lines, which attests to the effect of a gravitational field on the rate of clocks. All of these effects can be measured with much greater precision in space than on the surface of the Earth, permitting more accurate predictions of the gravitational field strength.

TESTS OF GENERAL RELATIVITY THEORY IN WEAK FIELDS

Deflection of Light

Currently, we can verify the predicted deflection of a light ray grazing the limit of the Sun with about 2 percent uncertainty. But we could improve this by 2 orders of magnitude if we could make the measurement with an optical interferometer flown on the Shuttle. This instrument would consist of an articulated pair of stellar interferometers, having their viewing axes approximately 90 degrees apart. It would have two pairs of mirrors 25 cm in diameter and an interferometer length of 2 m. A free-flying spacecraft could improve even on this precision by providing longer exposure and more stable pointing.

Gravitational Red Shift

The gravitational red shift is a consequence of the difference in the rate at which identical clocks measure time at different depths in a gravitational well. This effect has already been found to agree with the prediction of the theory of general relativity to within 1 part in 10^4. The experiment consisted of measuring the rate of a hydrogen maser clock as it was carried to a height of 10,000 km on a rocket. However, a qualitatively different test of general relativity theory could be performed by carrying an improved hydrogen maser close to the Sun, where the red shift will be more pronounced since the clock will be deeper in the gravity well. Significant variance of the measurements made there from the predictions of general relativity would cause a major rethinking of the theory.

Relativistic Frame Dragging

There is another prediction of the general theory that has never been tested. This is a nonstatic effect, and it states that rotating bodies drag nearby inertial frames. Although the effect is exceedingly small in weak fields near solar system bodies, it might be enormous and astrophysically important near a rotating black hole. The relativity gyroscope experiment called Gravity Probe B has been devised to search specifically for the frame-dragging effect produced by the rotating Earth. It will use the most precise gyroscopes yet devised. This mission has been likened in importance to the classical Michelson-Morley ether drift experiment of 1887. The proof that there was no ether drift buttresses Einstein's special theory of relativity and has changed fundamental concepts of space and time. Although many times more sophisticated than any experiment yet attempted in space, there is considerable confidence that the Gravity Probe B mission will be successful. Gravity Probe B should be flown before 1995 unless the consequences of the *Challenger* accident delay it.

PRINCIPLE OF EQUIVALENCE

General relativity is based on a fundamental principle called the principle of equivalence. The principle asserts that the gravitational mass of an object, that is, the quantity that measures the gravitational force it produces, is identical to the mass that responds inertially to any force. In short, it states that gravitational and inertial masses are equal. The validity of the principle has been demonstrated to a level of one part in 10^{11} in the famous Eotvos experiment. Shuttle flight of an experiment to test this equivalence at the level of one part in 10^{14} is proposed for the near term (before 1995). During the period covered by this study, a similar experiment flown in a free-flying spacecraft would provide a test to the level of one part in 10^{17}.

SECULAR CHANGE IN THE GRAVITATIONAL CONSTANT

Another important physical principle is called Mach's principle. It suggests that the expansion of the universe might cause the effective local value of the gravitational constant G to decrease with time as a consequence of the effect of distant mass on the inertial properties of local matter. Microwave ranging to a Mercury orbiter could improve our knowledge of the time rate of change of G by 3 orders of magnitude. A by-product of this experiment would establish the extent to which gravity is itself a source of gravitation.

GRAVITATIONAL WAVES

In Newtonian theory gravitation propagates instantaneously over infinite space. The concept of waves is not applicable. In contrast, Einstein's general relativity requires gravitation to propagate with the speed of light, just as does electromagnetic radiation. Electromagnetic waves jiggle charged particles; gravitational waves accelerate mass. When traversing a large object, a gravitational wave will deform it. In the language of relativity, a gravitational wave ripples the curvature of space-time, deforming any mass that sits in space.

The detection of gravitational waves is one of the most challenging problems in experimental gravitation today. Observation of gravitational waves would open new astronomical windows. It would provide information about exotic sources of gravitational radiation: collapsing stellar cores, colliding neutron stars or black holes, decaying binary star systems, and rotating or vibrating neutron stars. In the meantime, the discovery of a radio pulsar in a binary system containing, most likely, another neutron star has provided very convincing evidence that gravitational waves do indeed exist. The orbit of this system is decaying almost exactly as expected if such waves were being emitted.

If the explosive events in quasars and other active galaxies are generated by black holes or supermassive black holes, each explosion must generate a great gravitational wave that rattles everything in the universe. On the other hand, the radiation produced by many astronomical interactions, such as that of a black hole with neighboring matter, is of a very low frequency—below 10 Hz. Its detection requires an observatory in space, free from interference by seismic noise. A gravitational wave detector consisting of three spacecraft orbiting the Sun, each one a million kilometers from the next and possessing a precise system for monitoring their separation by laser ranging, would allow a detection of gravitational waves from astronomical sources in the range of periods from 0.3 s to 10 days. Gravitational waves would cause the distance between these spacecraft to oscillate. The estimated sensitivity achievable with such a system is one part in 10^{22} for narrow-band periodic signals and as much as one part in 10^{20} for transient pulses at megahertz frequencies. Such a detection system offers us our best chance of directly observing the radiation produced by distant matter accelerating in strong gravitational fields such as those produced by black holes.

Pulsars spinning with periods close to a millisecond approach relativistic instability; their surfaces move at close to the speed of light. The discovery of such objects could provide the frequency key to ground-based gravitational wave detectors in their search for gravitational wave radiation. The steering group recommends building a very large proportional-counter x-ray detector with a receiving area of about 100 m^2 that could be attached to the Space Station or orbit as a free flyer. This very large detector would search the sky for very fast x-ray pulsars.

PRE-1995 PROGRAM FOR RELATIVISTIC GRAVITATION

In summary, the steering group anticipates that several space experiments prior to 1995 will advance our understanding of general relativity in weak fields and offer a possibility of detecting gravitational radiation. These are:

1. The flight of Gravity Probe B;

2. Microwave ranging to the Galileo-Jupiter mission to search for low-frequency gravitational waves;

3. Microwave ranging to the Mars Observer spacecraft to improve the accuracy of measurements of the gravitational red shift, and variation of G with time;

4. Shuttle flight of a cryogenic experiment to test the weak principle of equivalence to one part in 10^{14}.

RECOMMENDED PROGRAM FOR RELATIVISTIC GRAVITATION: POST-1995

The major elements in the program recommended for the years 1995 to 2015 are:

1. *Laser Gravitational-wave Observatory in Space (LAGOS).* This mission will attempt to detect gravitational radiation at frequencies below 10 Hz from space. The mission, as proposed, consists of an optical heterodyne interferometer system accurately measuring the separation of three spacecraft in orbit.

2. *Mercury Relativity Satellite.* An improved measurement of the time rate of change of the gravitational coupling constant such as could be obtained by microwave ranging to a spacecraft orbiting Mercury.

3. *Precision Optical Interferometer in Space (POINTS).* This instrument will provide a second-order test of the effect of the Sun on electromagnetic radiation.

4. *STARPROBE.* This experiment involves the flight of an accurate clock (hydrogen maser) on a spacecraft close to the Sun, allowing the measurement of the gravitational red shift to the second order.

5. *Principle of Equivalence Experiment.* This experiment will be mounted on a free-flying spacecraft and will test this principle to one part in 10^{17}.

6. *Large-Area X-ray Detector.* The flight of such a detector with microsecond timing capability will allow detection of x-ray pulsars.

The successful implementation of this strategy should leave us with a very good understanding of the validity of the general theory of relativity in weak fields. It would also advance our knowledge of the behavior of matter in the neighborhood of objects such as black holes, where gravitational effects occur in fields far stronger than those hitherto observed.

B.
MICROGRAVITY SCIENCE

BACKGROUND

The microgravity environment of a space platform may provide a useful arena for testing basic theories of matter and observing new processes and new states in matter. Gravitational fields cause nonuniformities in the distribution of matter in a given sample and can cause fragile structures to collapse. The spacecraft environment can provide a very low effective gravitational field that might provide protection from these effects. Under conditions of low gravity, we may enhance our understanding of nonequilibrium phenomena in fluid flow, and in condensation, combustion, and similar dynamic processes. Low-gravity conditions may also allow the development of static or dynamic states of matter that cannot exist in normal gravitational fields.

OBSERVATION OF STATES IN EQUILIBRIUM

Three categories of investigations have been considered in these studies of states of equilibrium. The first deals with the case in which gravitational effects induce nonuniformity in the equilibrium state of a system, and thus prevent the observation of particular states of equilibrium, such as phase transitions near critical points. These states involve correlation lengths that are long compared with the distance over which uniformity in a system can be maintained in normal gravitational fields. The most well-known example is the continuous phase transition in liquid helium at its lambda point. Plans are well advanced to carry out an experiment investigating this phenomenon in space, where gravitational effects will be small enough to allow uniform temperature in an extended sample of liquid helium. This experiment should be completed before 1995.

OBSERVATION OF STATES DESTROYED BY GRAVITY

Another category of investigation involves the study of stationary states of matter that gravity destroys rather than distorts. For example, there is the possibility that in low gravity, objects can develop so-called fractal aggregates. These structures may be so fragile that they can exist only in a microgravity environment. In another case, gravitational effects can interfere with the evolution of a precipitate because of flows induced by buoyancy or because of sedimentation. In a microgravity environment these effects could be avoided, and precipitation solely under the control of diffusion could be observed.

STATES FAR FROM EQUILIBRIUM

A third class of phenomena that can be observed only under conditions of low gravity are those that exhibit complex dynamical behavior as they are driven far from equilibrium. Plans have been formulated for studying examples of this sort of behavior on the Space Shuttle, including the combustion of clouds of particulates, or surface-tension-driven hydrodynamical flows. But the steering group believes that the possibilities for research in this field are greater than we now realize and that they may have important implications for biology. Many questions beckon for answers: Will a given process will be chaotic or not? Will spatial patterns formed be stable? What is the role of the gravitationally induced breaking of underlying symmetries, such as the front-to-back symmetry in flames?

CONCLUSIONS AND RECOMMENDATIONS FOR MICROGRAVITY SCIENCE

In its treatment of microgravity science here, the steering group has concerned itself solely with basic scientific questions. *Until these are answered, there does not seem to be any way to structure a rational program of materials processing in space.* A basic research program of this sort is a necessary precondition to the development of an applied program. As a branch of space science, microgravity science is in its infancy. Thus, before we can gauge the prospects of the field over the next 20 years, *we must know the results of preliminary experiments now being developed.*

If the nation hopes to attract expert scientific talent into the field, flight of these experiments merits very high priority.

The following are specific recommendations regarding microgravity studies:

- In scheduling experiments for flight, NASA should make every effort to fly the best of the microgravity physics and chemistry experiments as soon as possible, and see to it that the results are rapidly published.
- Spacecraft gravity levels and vibration spectra should be precisely measured, characterized, and displayed on those spacecraft carrying chemistry and physics experiments.
- Strategies for producing the lowest possible gravity conditions should be considered at this time, since experiments dealing with long-range order are open-ended in their low-gravity needs.

7
Life Sciences

BACKGROUND

The study of life on Earth ranges from elucidating the evolution of the earliest self-replicating nucleic acids to describing a global ecology comprising over three million species, including humans. Though life has shown enormous diversity and complexity over the last 3.5 billion years, its unifying principles are becoming ever clearer.

Chemical and fossil evidence show that earth life as we know it today evolved by natural selection from a few simple cells called prokaryotes because they lacked nuclei. The earliest prokaryotes probably already had mechanisms that allowed them to replicate their genetic information, encoded in nucleic acids, and to express this information by translation into various proteins. These first cells were somehow formed during the Hadean eon that spanned the epoch from about 4.5 to 3.5 billion years ago. Analyses of the chemical compositions and reactions occurring on other planets, on comets and asteroids, and in interstellar space help us reconstruct the steps leading to the formation of the organic building blocks of biopolymers. Understanding this prebiotic evolution is one of the major goals of the *exobiology* program—biology of the early Earth and elsewhere in the universe.

A second goal of exobiology is to understand the evolution of the first cells with true nuclei. These cells, called eukaryotes, were the precursors of "higher" organisms: the unicellular protists, fungi, plants, and animals we know today. The important organelles of energy metabolism—plastids and mitochondria—originated 2.0 to 1.5 billion years ago by the symbiosis of prokaryotes. In this process bacteria having one set of specialized functions were engulfed by host cells with complementary requirements and functions.

By the early Archean, more than 2.2 billion years ago, the biota had used the process of photosynthesis to create an oxidizing atmosphere from one previously poor in oxygen. Carbon dioxide was also removed from the atmosphere in the form of carbonate precipitates. Myriad bacteria, molluscs, corals, and other organisms contributed to vast limestone deposits and continue to do so today. With these and other processes Earth's biota have transformed a sterile planet, intermediate in character between Venus and Mars, into the living planet we now enjoy.

Global biology concerns itself with the cumulative changes wrought by the biosphere on the atmosphere, hydrosphere, and geosphere. But, conversely, the physical and chemical constraints of Earth have acted on the biota as well, and this too falls in the province of global biology. It is hardly a coincidence that Lyell's *Principles of Geology* was so influential in the development of Darwin's early interest in the relationship of life and Earth. As the impact of human activities, such as burning fossil fuels and deforestation, increases, the predictive power of global biology will become as important as its ability to interpret the history of Earth.

Global biology and exobiology address questions about the nature of Earth and the origin of life that have concerned us since even before the writing of Genesis. Yet, as these questions are now formulated, both global biology and exobiology are young sciences. As they have grown, they have profited from space missions and from collaborations with earth scientists and planetary scientists. As a result, the goals of these fields are now well defined; we can proceed with specific missions and programs confident of valuable results.

In contrast to global biology, exobiology, and the other space sciences, the fields of space biology and space medicine are still formulating their basic theories and research goals. A significant proportion of the entire scientific research effort of the United States addresses questions closely relevant to space biology and space medicine. Yet there have been so few space flights committed to biological observations that the major challenge facing space biology today is to determine how valuable a low gravitational field (micro-g) laboratory might prove in addressing fundamental research problems in biology.

Some single cells must be able to detect small changes in the magnitude and direction of gravitational force. At a fundamental level we need to understand the molecular mechanisms whereby a cell detects gravity and converts this signal to a neuronal, ionic, or hormonal response. Plants and animals that have evolved on Earth have done so at one g; they respond to gravity. For example, plant roots grow down and shoots up—gravitropism. Fertilized eggs orient their cleavage planes relative to the gravity vector. In larger animals and plants many of the responses to gravity are additive. Hydrostatic pressure and muscle tension are good examples: the impact of gravity is probably greater on a giraffe than on a bacterium. Studying systemic effects produced by variations in gravitational force will not only contribute to our theories of physiology but test their predictive value as well.

The steering group anticipates that manned space flights extending over years will pose severe psychological and physiological problems. The initial concern of space medicine is primarily to identify and characterize those physiological systems that do not adapt well to space flight or that do not subsequently readapt to one g. The vestibular and the cardiovascular systems, for example, may adapt and readapt without further intervention. In contrast, extrapolation from very limited observations indicates that the effects of bone remodeling will grow more severe with passing months. Similarly, shielding cannot easily control cosmic radiation consisting of the nuclei of heavy atoms (so-called heavy ion radiation). The impact of one such particle can cause the death of nondividing cells, such as brain cells. It would be unconscienable to launch long-term manned space flights without a much better understanding of these phenomena and, where indicated, development of appropriate prophylaxis and treatment. Until these measures have been taken, it is questionable whether such flights should be planned or contemplated. With the number of Shuttle flights now projected to study these phenomena, we would still have only a meager base of data by 1995. Hence, of necessity, our projections for the period from 1995 to 2015 are tentative.

While priorities may change with time, present observations and extrapolations indicate clear research goals of both basic and clinical significance. Realization of these goals demands the deployment of a dedicated life sciences laboratory equipped with a vivarium and a centrifuge of at least an 8-foot radius on a space station. Nothing less can allow us to determine whether microgravity provides an important tool for the study of biology; nothing less will give us our best chance to assure the health, performance, and welfare of astronauts on long-term missions.

LIFE SCIENCE GOALS AND MAJOR QUESTIONS

Exobiology Goals

The goal of exobiology is to understand the origin and early evolution of life and its cosmic distribution.

Complex anaerobic ecosystems fueled by photosynthesis existed at least 3.5 billion years ago and possibly earlier. They had produced an oxidizing atmosphere by 2.0 billion years ago. The oldest fossils of eukaryotic organisms are found in rocks about 1.4 billion years old. Exobiologists believe that the basic chemical components of the first cells had to accomplish two fundamental tasks: reproduction and energy storage. Both of these probably involved the replication of polymers of RNA. These first RNAs may have been catalytic, somehow directing polymerization of amino acids into proteins. Selective advantage would have accrued to those organisms that let them extract usable energy from organic molecules, sunlight, or minerals not in their final oxidation state. However, relatively complex, prebiotic organic reactions occurred not only on Earth. They also occurred on other planets and in interstellar space. The surfaces of clays or of interstellar grains may have catalyzed these early reactions, even though these same pathways are no longer used by existing cells. A study of this extraterrestrial organic chemistry should provide precedents for early reactions on Earth. It might also indicate whether conditions conducive to life have ever existed on other planets in our solar system or beyond.

Major Questions for Exobiology: 1995 to 2015

From 1995 to 2015 the steering group recommends addressing five major questions:

1. Are there organic molecules beneath the polar caps and in the sediments of Mars, in the atmospheres of Jupiter and Titan, in comets and meteorites, and in interstellar space? If so, what are they, and what is their distribution? More than 50 organic compounds containing up to 11 atoms have been identified in interstellar space; their relative concentrations differ enormously from that predicted by equilibrium thermodynamics.
2. How can we interpret the early fossils and deposition of material of biological origin on Earth and relate them to materials from other planets? Are there fossils or definitive evidence of liquid water or former life on Mars, and, if so, what role have they played in early biological evolution?
3. What models of prebiotic chemical reactions can be demonstrated experimentally? The correspondence with reactions that do occur or have occurred in nature should be established.
4. What models can be developed for the first replicating system and the first true cells? A critical evaluation of such models might outline the environmental characteristics required for the origin and evolution of life.
5. What is the phylogeny of the archaebacteria, eubacteria, and early eukaryotes, and what were the major events in the evolution of the eukaryotes?

Global Biology Goals

The goal of global biology is to understand the evolution of the biota and its interactions with Earth.

For nearly 4 billion years the biosphere has influenced, and been influenced by, the atmosphere, the hydrosphere, and the geosphere. The composition of our atmosphere, the abundance of water, the temperature, and the vast carbonate and ferric iron deposits are radically different from conditions on Mars and on Venus. Life has evolved under the constraints of available elements and energy fluxes of our planet. This is a dynamic system; the flow of solar energy to a rotating Earth, coupled with its own internal radioactive heat source, generates numerous cycles and changes, which act on widely varying time scales. Photosynthetic cycles, for example, have daily periods; while time scales for plate tectonics are in the millions of years.

Major Questions for Global Biology: 1995 to 2015

From 1995 to 2015 the steering group recommends addressing six questions:

1. What are the major features of the water cycle? Not only is water essential to life, it drives or influences nearly all the other biogeochemical cycles as a solvent, as a source of aerosols, or as a carrier of particulates. On a longer time scale it is the major agent of erosion and subsequent sedimentation.

2. What are the reactions and the fluxes of the other biogenic elements—carbon, nitrogen, sulfur, and phosphorus? During the past century the atmospheric concentration of carbon dioxide has increased from 200 to 300 ppm; methane, a trace gas, may provide a significant component of the flux of carbon. The turnover of atmospheric nitrogen is slow; we do not understand whether the seeming balance between nitrogen fixation and denitrification is affected by the production of nitrogen dioxide by lightning. Prior to industrialization most sulfur was introduced into the atmosphere by vulcanism. There is also significant biogenic production of hydrogen sulfide and $(CH_3)_2S$ by anaerobic organisms. Phosphates are dissolved or borne as particulates; animal depositions may account for most soil replenishment. It is far more difficult to measure flux than concentration of these compounds; yet values for flux are essential for proper modeling.

3. What are the major sedimentation and erosion processes that the biota influence? About 99 percent of Earth's carbon exists as carbonate sediments, most of them derived from biomineralization. Rates of erosion by wind and water depend on the extent and type of vegetation.

4. How productive are the major ecosystems? The rate of photosynthesis is influenced by the amount of light of appropriate wavelength reaching the Earth's surface, and, hence, is influenced by the Earth's albedo. Most of the energy converted by plants, algae, and photosynthetic bacteria is expressed in the synthesis of carbohydrates: $CO_2 + H_2O \rightarrow CH_2O + O_2$. The albedo of the Earth depends not only on its atmospheric composition, but also on the life in its oceans and on its land masses.

5. What are the impacts of major human activities? Extensive use of fossil fuels and other industrial activities are increasing atmospheric concentrations of carbon dioxide, sulfur dioxide, and other gases. Deforestation in the tropics leads to increased erosion, reduced synthesis of ozone by photosynthesis, and reduced transpiration of water. The release of chlorofluorocarbons into the atmosphere may be reducing the ozone layer and increasing the greenhouse effect.

6. Can the vast amounts of empirical data that will come from satellite and other global arrays of instruments be incorporated into a predictive model for Earth as a whole, leading to the formulation of new principles? To model Earth systems on time scales from days to millennia presents complexity without precedent.

Space Biology Goals

The goal of space biology is to determine whether the unique opportunity of experimentation at microgravity can advance our understanding of basic phenomena in biology.

Throughout its entire evolution, life on Earth has experienced only a one-g environment. The influence of this omnipresent force is not well understood except that there is clearly a biological response to gravity in the structure and functioning of living organisms. Access to a microgravity (micro-g) space station laboratory may facilitate research on the cellular and molecular mechanisms involved in sensing an acceleration as low as 10^{-6} g and subsequently transducing this signal to a neural, ionic, or hormonal signal. Propagating selected species of plants and animals through several generations at micro-g in such a space laboratory would advance our understanding of these biological responses.

Major Questions for Space Biology: 1995 to 2015

From 1995 to 2015 the steering group recommends addressing four major questions:

1. How do plant cells detect the gravity vector and transduce this force to hormonal and nonhormonal signals? This gravitropic response utilizes growth-stimulating hormones, such as gibberellin and indoleacetic acid, and the inhibiting hormone abscisic acid. These same hormones, along with electric current, are involved

in phototropism; the interactions between the two responses are complex. Investigations of these processes are important for the development of a successful Controlled Ecological Life Support System (CELSS).

2. Can higher plants and animals be propagated through several generations at micro-g? Although many embryos orient their cleavage planes relative to the gravity vector, we do not understand whether gravity, per se, is essential to gametogenesis, fertilization, implantation in animals, organogenesis, or development of normal sensorimotor responses. Given the effects of micro-g on demineralization in bones, muscle wasting, and vestibular function, there is some question whether vertebrates can develop normally at micro-g.

3. What is the relative contribution of gravity to sensorimotor functions? The otolith organs in the inner ear allow us to detect linear accelerations in three orthogonal directions and distinguish these from rotation, which is detected by the semicircular canals. These responses, already complex, are further integrated with visual and proprioceptive input. The ability to remove or to vary signals from the otolith should help us understand the interactions of these sensory systems and shed insight on the nature of motion sickness.

4. What are the fundamental biochemistry and physics of biomineralization? Understanding the physical and chemical processes of biomineralization on Earth is necessary to fully understand the potential effects in microgravity environments. This process almost always occurs within membrane vesicles or is associated with polymers of carbohydrates or of proteins. Scores of biominerals exist, the most common being $CaCO_3$, calcite or aragonite, and $Ca_{10}(PO_4)_6(OH)_2$—the hydroxyapatite of bones. Macroscopic solubility products do not readily explain the growth of these aggregates of crystallites.

Space Medicine Goal

The goal of space medicine is to understand the human biology underlying the prophylaxis and therapy of maladaptations encountered in extended space travel and to develop prophylaxis and therapy to treat them if it is feasible.

Space flights extending for weeks or months have already

caused several physiological problems that could endanger astronauts during longer flights or on their return to Earth. At micro-g, fluid accumulates in the upper body; subsequently, astronauts must remain horizontal for hours to weeks on return to Earth. During flight, extensor muscles atrophy, and, of greater concern, bone calcium and phosphate are lost from weight-bearing bones. We do not know whether these effects reach a plateau. If not, they could irreversibly compromise the health of an astronaut or even lead to death. Numerous other effects of prolonged weightlessness, compounded by the rigors of the confined environment of a station, pose serious threats to health and performance. The heavy ion radiation of outer space is not only mutagenic but also could have disastrous effects on the brain. Each such particle inflicts severe damage or even death to nondividing cells.

Major Questions for Space Medicine: 1995 to 2015

From 1995 to 2015 the steering group recommends addressing six major questions:

1. Are there adverse effects of weightlessness that grow progressively more debilitating as flights are extended incrementally from months to years? We already anticipate severe problems with bone loss, and probably with muscle atrophy and cardiovascular deconditioning. Others that cannot be predicted may also occur. These could involve the immune and the endocrine systems. The rigors of confinement will probably compound these physiological effects and profoundly affect behavior. A series of controlled experiments will help to sort out the effect of inevitable human variability on responses.
2. What are the hormonal, nutritional, and mechanical correlates and mechanisms of biomineralization? Although doctors may develop empirical procedures to alleviate the severity of demineralization, we must understand the underlying molecular and cellular biology of bone remodeling and of calcium and phosphate homeostasis. A similar question concerning biomineralization was posed for space biology, reflecting the frequent and productive interactions between clinicians and biologists.
3. What are the fundamental genetic, hormonal, and mechanical factors that determine muscle development and maintenance? Our working hypothesis is that during muscle wasting the rate of breakdown of muscle protein exceeds the rate of synthesis. We

must understand the factors coupling weightlessness to the controls of those genes encoding muscle proteins as well as those genes encoding muscle-specific proteases.

4. What are the biological effects of prolonged exposure to heavy ion radiation? It is difficult to reproduce with particle accelerators the exact spectrum of radiation encountered in outer space where heavy iron, ^{56}Fe, traveling at 0.9 times the speed of light, is the predominant particle. Cultures of cells as well as plants, rodents, and higher primates must be exposed and subsequently analyzed to enlarge the empirical data base. In parallel, we should explore the basic biophysics of the heavy ion damage in hopes of developing prophylaxis and therapies.

It would be imprudent even to plan extended missions until these serious medical issues are resolved.

5. How can we ensure adequate life support systems for long-term space travel? The development of a Controlled Ecological Life Support System (CELSS) is essential to missions of long duration. Although this development is, in a sense, a technological issue, it is so complex and requires so many advances in our understanding of biology that we can regard it as a major scientific goal that spans the interests of space medicine, space biology, and global biology. Aside from the challenges of growing plants at micro-g, and the utility of CELSS as a life support aid, the concept of constructing, at least partially, an artificial ecosystem is of great interest to the science of ecology. It may offer a research tool of considerable value for study of the principles by which natural ecosystems function.

6. Under the conditions of space, what are the optimal interactions between humans, machines, and computers? Many construction and observation tasks will be done under extremes of pressure, temperature, and radiation hostile to the human organism. Human judgment and ingenuity are valuable or indispensable to some of these jobs. This should not be seen as a choice between man and robot, but as a challenge to integrate their respective capabilities. This will require fundamental research not only in machine design but also in human neurophysiology.

RECOMMENDED PROGRAM: POST-1995

Two basic requirements dominate most of these recommendations. Global biology and exobiology require ground-based observations and experiments to calibrate and verify satellite observations. Space biology and medicine require one-g controls for experiments and observations at micro-g.

Recommendations for Exobiology

The steering group envisions a series of observation and sample return missions complemented by ground-based analyses and experiments.

1. Determine the properties of the atmospheres of Mars, Titan, and Jupiter with greater precision. The reactions leading to any organic compounds found there should be compared with models for chemical synthesis on the primitive Earth.
2. Return sedimentary rock and soil samples from Mars after a careful analysis and selection of sites. The initial compositional analyses can be done at the sites. Complete analyses will require a full range of laboratory procedures including electron microscopy and microprobe analyses. This study should include searches for organic molecules and for fossil evidence of primitive life forms, and determination of important isotopic abundances, such as carbon, nitrogen, oxygen, and hydrogen.
3. Return samples from asteroids and comets to Earth and analyze them with the methods now used to analyze meteorites and captured interplanetary dust particles. The design and execution of these missions should be closely coordinated with planetary sciences.
4. Continue to explore fossils, isotope enrichment, and other evidence of prebiotic and early life on Earth.
5. Refine laboratory experiments simulating likely prebiotic reactions as more results from missions accumulate and as molecular biology lends more insights into the fundamental structures and reactions of cells.

Recommendations for Global Biology

The steering group projects a series of satellite flights to define the major ecosystems over the entire Earth and to measure remotely their contributions to the major biogeochemical cycles. Data obtained from these flights should all be calibrated against measurements at selected sites to assure detection of long-term trends.

1. Determine the predominant plants for the major ecosystems, and measure their chlorophyll content and rate of biomass formation.

2. Measure the seasonal variations in carbon dioxide fixation for these ecosystems.

3. Measure the production rates, reactions, and fluxes of several trace gases for the major ecosystems. These measurements will require the refinement of gas chromatographs, mass spectrometers, color imagers, laser fluoroscopes, and synthetic aperture radar.

4. Compare the impacts of human activities, such as deforestation and industrialization, with well-established baselines. These goals are included in the Mission to Planet Earth (described in Chapter 2). The two programs should proceed in close collaboration.

5. Develop correlative models to accommodate vast amounts of diverse data. This will require that special attention be paid to data reduction and archiving and to computational facilities. These correlations should lead to interpretative and predictive models.

Recommendations for Space Biology and Space Medicine

The steering group believes that these disciplines require a series of missions, each addressing one of the major research problems previously discussed.

1. Construct a dedicated life sciences laboratory with a large variable-speed centrifuge to hold plants and animals and to provide one-g controls. Without adequate controls, most of the experiments at micro-g will be of limited value. This centrifuge will also facilitate experiments addressing the effects of reduced gravity as found on the Moon. (As will be required for micro-g experiments in the physics of solutions, some experiments treating the settling of amyloplasts or growth of crystals will require a frequency analysis and isolation from accelerations exceeding 10^{-5} g.)

2. Consistently employ four to eight critically selected plant and animal species, such as the nematode *Caenorhabditis elegans* and the cress *Arabidopsis thaliana* for as many gravity-related experiments as possible. This focus will extend and refine descriptions of the physiologies and genetics of these species, including techniques to clone them and extensive analyses of their genomes.

3. Precede micro-g experiments, which will be infrequent and expensive, with thorough simulations at one-g (e.g., extended bed-rest for cardiovascular deconditioning or demineralization and clinostat experiments for gravitropism). In-flight controls at one-g are mandatory to account for potential artifacts, such as the acceleration of launch, vibrations, and astronaut activities.

4. Evaluate the biological effects of heavy ion radiation thoroughly using accelerator sources. If possible, these experiments should employ the selected organisms already mentioned.

5. Require a prototype of CELSS to function satisfactorily on Earth for several years before attempting to launch such a system into space. Evaluate in flight those components whose functions might be altered at micro-g.

6. Establish empirically a data base of the ranges of human physiological and behavioral responses to micro-g under prolonged isolation. All astronauts should participate in noninvasive or minimally invasive tests as well as monitors of behavior—activity levels and general speech patterns, for example. All such tests must be consistent with well-established standards of personal privacy and medical ethics and must not interfere with astronaut safety or job performance. Many of these basic data—such as pulse, blood pressure, basic metabolic rate—can be obtained from miniaturized devices worn on the skin, or from simple automated analyses of urine and drops of blood. (Although this study does not address the delivery of health care, it is obvious that many of the measurements required for research are required for diagnosis. Economy can be realized without compromising either function.)

7. NASA should make a major investment in robotics, both developing new instruments and computers and optimizing the interactions between humans and robots.

Significance of Recommendations

By 2015 we hope to have an inventory of organic compounds found on other bodies in the solar system. By 2015 we expect more definitive information on the past and present liquid water on Mars, as well as the results of the preliminary search for fossils and other evidence of past life.

The greatly enlarged data base of global biology and earth science should permit refined predictions of atmospheric patterns over decades. We should have much more insight into the influence of life on the general evolution of Earth. In particular, we should understand whether major perturbations, such as glaciation or mass extinction, reflect events originating outside the Earth or if they are predictable transitions in the evolution of a closed system with constant energy flux. The knowledge gained from these measurements will focus our searches for minerals and fuels. Also, with an appreciation of the ecological impacts of human activities, rational and balanced conservation programs can be implemented. It is quite possible that an increased respect for our planet will accompany an increased understanding of it.

The steering group has proposed investigations at micro-g of several biological phenomena, such as gravitropism development, sensorimotor integration, and bone remodeling. It has also called for a Controlled Ecological Life Support System. These initial experiments should determine whether micro-g offers an effective experimental approach to some basic problems in biology. If such experiments are performed with flexible formats and appropriate controls, they may well reveal unanticipated phenomena of even greater interest. Although serendipity is hardly the basis of a research strategy, the steering group believes it a wise investment to apply some of the resources of the dedicated life sciences laboratory to define the full range of questions best addressed at micro-g.

By 2015, and even by 2005 if there is a dedicated life sciences laboratory, we should have fully described the physiological and behavioral responses of human beings subjected to micro-g and to heavy ion radiation for periods of over a year. Anticipating significant adverse effects, we should at least have defined which of these effects might be alleviated and be well on the way to realizing treatments. We should have acquired a greatly refined insight into the unique limitations and capabilities of humans and of robots in space flight. Most importantly, understanding the optimum interactions of humans and robots should permit us to evaluate rationally their appropriate roles in commercialization, exploration, and research in space.

This program in life sciences should be adequately funded and properly integrated with the other space sciences and with ground-based investigations. It will make a significant contribution to understanding the evolution of our planet and ourselves. If pursued with reason and caution, it will help define our proper relationship to Earth.

8
Interdisciplinary Studies

Before this study began it was already clear that there would be some overlapping interests among the various disciplinary task groups. From the beginning of the study, the task groups were urged to join forces with each other in identifying problems of common interest, either intrinsically or because they required common means for investigation. This approach proved fruitful. The task groups in astronomy and astrophysics and in planetary and lunar exploration recognized their common interest in understanding the processes by which planetary systems develop. Both groups, therefore, emphasize developing techniques to observe such systems as they are being born and when they are mature. The astronomical and the relativity disciplines both appreciate the potential significance to cosmology that objects such as black holes or cosmic strings may have. Additionally, detection of gravitational radiation is important to the field of relativity theory and, as another means of understanding fundamental cosmological processes, to astronomy as well.

Solar studies were another area where interdisciplinary links were evident. The Sun is an object of great interest to astronomers as well as to solar and space plasma physicists. Likewise plasmas, representing the dominant form of matter in the universe, are of intense interest to solar scientists, solar system physicists, and astronomers alike. Planetary scientists, solar and space plasma physicists, and relativists all have requirements for spacecraft that orbit Mercury and that approach within a few solar radii of the Sun. This common interest accords a higher priority to such missions.

Earth scientists and life scientists alike could have designed the Mission to Planet Earth. The configuration of that mission has been determined by contributions from the earth science and life science task groups. As well, life scientists share with planetary scientists an interest in understanding how life developed in this solar system and whether or not it exists elsewhere. It is not possible to understand Earth by studying it in isolation and out of its context as one of the terrestrial planets, each of which has followed a different evolutionary track.

This list of interdisciplinary projects is not exhaustive. However, it should demonstrate that a balanced space science program in which multidisciplinary investigation can flourish should be maintained during the next century.

9
Human Presence in Space

SPACE AGE SCIENCE

It is difficult to determine which of the scientific projects comtemplated in this study, other than those in space medicine, compel the presence of humans in space. In fact, there may be no others. With sufficient resources, we might devise automated systems that could substitute for people, performing all of the necessary functions usually associated with humans. People would, in turn, control the machines from Earth. On the other hand, it appears that under certain circumstances people are able to function productively in space and perform tasks in support of scientific investigations. At present, we lack enough information to judge where the balance between manned and unmanned missions should lie.

Some space science missions at the beginning of the twenty-first century may be intended to pave the way for the expansion of humanity into deep space. For many reasons, not all scientific, human activities may extend into an increasing arena of space. With the advent of space stations, plans are already being made by a number of national space agencies for the continuous presence of men and women in low earth orbit, beginning in the

1990s. Such activities in the next century may extend to geosynchronous orbit, and possibly to the regions of the L4 and L5 points. The National Commission on Space has also expressed interest in establishing inhabited stations on the Moon and Mars. Space science practiced at the frontier requires a wide variety of innovations in observational and control capabilities, instrumentation, and propulsion methods. Thus, the pursuit of space science and its supporting functions should be a powerful driver of advanced technology, extending the capabilities of unmanned spacecraft. Advances in the technology of sensors, robotics, artificial intelligence, and parallel computation may enable the development of a new generation of autonomous decision-making machines that will extend exploration and intensive study into remote parts of the solar system—and eventually beyond—without a human presence. Earth orbit can become a proving ground for the deployment of robots, automated observatories, and advanced data management systems.

THE SCOPE OF HUMAN PRESENCE IN SPACE

The space stations of the United States and the Soviet Union are the first steps toward a sustained human presence in space. It is impossible now to predict either the pace or the ultimate extent of this expansion into space. The human-inhabited sphere may never extend beyond low earth orbit. Whether its boundaries are near Earth, on the surface of Mars, or somewhere else, this human-inhabited sphere will be the base from which many future space science investigations are conducted. Conversely, these investigations will provide the foundation needed for continued expansion of this sphere, if called for. Space science experiments, tended in space by human beings, may provide the most important rationale for the staging, assembly, maintenance, repair, and operation of major space facilities (e.g., space astronomical telescopes, earth science experiment payloads/platforms, launch vehicles for planetary missions).

The steering group expects that the sphere of human presence in space will have relatively distinct boundaries. Within this sphere human presence will be pervasive and well-supported. Many scientific investigations will be carried out under direct human supervision, much as they are on the ground; others will be conducted in a largely automatic mode, with general supervision from scientists on Earth or perhaps elsewhere within the inhabited sphere.

This confinement of human activities to regions where they can draw upon a host of well-established facilities is advisable for two reasons. First, the capability of humans to make judgments is optimized when there is an opportunity for adaptation, over a long period of time, to the new environment. Second, human manipulative and observational skills can rarely be used effectively without the support of a large array of sophisticated instruments, machines, and facilities. Neither of these two conditions is generally met by brief forays of human beings into regions far from the facilities that support their sustained presence. This applies to manned excursions to Mars, for example, if the human-inhabited sphere is restricted to space near the Earth.

Further, it is important to recognize that the limitations on human survival in space are not well known. At present, we are not certain that mission times can be extended greatly beyond those already experienced, even with considerable technological progress. Low gravity leads to loss of bone mass and other physical effects. High-energy, heavy ion radiation causes irreversible damage to cells, including brain cells. Human relationships in a small, isolated group can badly deteriorate and lead to the loss of functional capabilities. We have not demonstrated the feasibility of a closed ecological system yet, and resupply at a great distance for a long period could be formidable. We must address these issues before we can reach a final decision about the nature and extent of human involvement in expanding the frontier of space.

10
International Cooperation

Specific issues involving international cooperation are treated in the various task group reports. The steering group endorses these treatments. Here the steering group wishes to deal with some principles common to all disciplines of space science.

Space science is now an international activity. More than a decade ago, the United States clearly dominated space science. That is no longer the case. The American space science program is still preeminent, measured in terms of missions ready for launch or being prepared. But the present crisis in launch capability has crippled the American program. Measured in terms of the number and quality of missions actually being launched, the Soviet Union is now the leader in space science. The European program, while substantially smaller than the United States and Soviet programs, has grown to major proportions; Japan's program is also substantial. All of these space science programs are carried out by agencies that have access to both earth-orbital and deep-space launch vehicles, although not all are equally capable in this respect. With four fully independent space science programs now in existence, it is time to consider the advantages of expanding joint participation in some scientific projects.

Cooperation can take a variety of forms. The simplest level entails coordinating programs at the planning stages or exchanging data after they have been collected. At a more significant level, scientists can participate in projects planned and executed by space agencies of other nations. Other cooperative modes include joint investigations involving many spacecraft from different communities. Examples include the International Solar Terrestrial Program (ISTP) and the orbiting very-long-baseline interferometry system for radio astronomy, known as QUASAT. A future venture of this type might be the establishment of an extensive network of ground stations coordinated with orbiting vehicles. The most ambitious level of international cooperation involves joint planning and implementation of cooperative projects.

In carrying out the scientific programs recommended in this report, the steering group believes that international cooperation, at all the levels listed above, should be considered. From a scientific point of view there can be numerous advantages to international cooperation. By coordinating or combining resources—intellectual, technological, and economic—scientific advances will proceed faster and more efficiently. While national imperatives other than the advance of science may play a role in motivating international cooperation, *it is important that the scientific goals be held in a primary position. The impact of international cooperation on the direction and balance of the national space science program should be carefully considered and evaluated.*

Several factors should be considered in formulating the varied approaches to international cooperation the nation will require. Among the most important considerations are the past history of cooperation, and the existing level of communication among the potential partners on both scientific and technical matters. Further, the quality and anticipated stability of the political relationship between participating nations should be evaluated. *It is essential that the structure of cooperative programs should be robust and resistant to disruption by unanticipated changes in the relationship that may be imposed for reasons that are outside of the scientific programs.*

It is important to ensure that all sides in any cooperation should obtain reciprocal advantages and should perceive them to be so. Thus, cooperation should be arranged to take advantage of mutually complementary capabilities. Therefore, international cooperation will be most productive if the separate partners have strong independent programs. In these cases cooperative ventures would be seen as a means to enhance these strong independent programs, and not as an essential strategy for the conduct of space science. Certainly, other forms of cooperation, involving nations that do not have their own means of access to space, should by no means be excluded. On the contrary, they should be encouraged.

The advantages of international cooperation are inevitably accompanied by operational and financial burdens, no matter how cleanly the technical interactions are arranged. The larger the number of partners involved, the larger will be these imposed burdens. In general, the complications and burdens of a cooperative program will be minimized if there are fewer major partners. *Thus the steering group recommends that large cooperative projects be held to only two major partners, at least until a record of successful experience accumulates.* However, where cooperation of more than two partners offers advantages, the complications and burdens should be recognized and taken into account.

In order to implement desirable levels of cooperation, two essential steps should be taken. *First, the United States should establish a national policy with respect to the goals of international cooperation in space science.* This policy should be guided by a primary commitment to enhance the scientific returns of cooperative ventures and should establish guidelines to ensure feasibility, maximize productivity, and minimize costs of these ventures. *Second, it is essential to establish suitable mechanisms for planning and implementing the various kinds of cooperative endeavors.* The most ambitious approaches to cooperation will require implementation of agreements at the highest national levels.

11
Preconditions and Infrastructure

Many developments in technology–some of them extremely challenging–will be needed if the program recommended here is to be successful. Specific needs for each discipline are identified and discussed in the individual task group reports. The steering group endorses those recommendations.

This study focuses on large-scale scientific undertakings. These cannot be implemented unless a certain set of preconditions, listed below, are satisfied.

- *These undertakings must be built on a solid foundation of supporting research and technology. This foundation must include vigorous theoretical and ground-based laboratory studies.* Scientific progress does not begin and end with the construction of flight hardware and the acquisition of data. Nor is it sufficient to confine theoretical analysis or laboratory support to preparation for specific missions or the interpretation of mission results. Theoretical and laboratory studies establish the framework in which data can be understood; they are not captives of specific missions, nor can they be started and stopped at will. Stable funding for these supporting activities will pay off handsomely.
- *For very similar reasons, small-scale, exploratory flight activities such as the present Explorer, Observer, Spartan, and subor-*

bital programs must be allowed to flourish. The steering group foresees no qualitative change in the way progress is made in science. Thus, we will continue to require missions with short implementation times. These missions may be designed to answer questions of detail, exploit findings of larger projects, or attack targets of special opportunity. Furthermore, participation in space science by graduate students, postdoctoral fellows, and young university faculty members requires projects that can be completed in one to three years. *The steering group believes that progress in space science critically depends on the full participation of universities because that is where much of the reservoir of scientific talent resides. The university space science community functions best in a program that includes short-term, small-scale projects. Thus, in order to foster space science and to ensure the viability and participation of that community, the space science program should be structured to include such small-scale projects with ready access to space flight.*

- *Balance in the research program must be maintained among groups at universities, industrial laboratories, and government centers in conducting space research.*
- *The laboratories used for space research should be amply furnished with state-of-the-art equipment.* Currently, the condition of equipment in most university laboratories can only be described as abysmal.
- A generic requirement in most of the fields covered by this study is for detectors that are more advanced than those now available. *Advanced programs for detector technology should be established and nurtured.*
- Computational facilities play an essential role in gathering, storing, and analyzing data. They also enable theorists to develop models and to test their models against experimental and observational data. *Computer facilities in the space program must be maintained at a state-of-the-art level, with regard to both hardware and software.*
- *As this report is being written, the nation's access to space has been severely curtailed. This situation accents the need for a sturdy, redundant system of acquiring access to space. Launch systems, delivery mechanisms, space platforms, and other such developments should never be looked upon as ends in themselves. Rather, they should be treated as tools to support well-defined objectives.* The science objectives have been the subject of this report.

If the conditions set forth in this chapter are met, the steering group is confident that the nation will be prepared at the turn of the century to embark on the scientific program recommended in this report, and that the future of the national program in space will be as bright as its past. Some of the participants in this study were present when the space age in science began at White Sands 40 years ago. The program designed here, along with the achievements of the past four decades, is a legacy they leave to their children.